LAB ON THE WEB

LAB ON THE WEB
Running Real Electronics Experiments via the Internet

Tor A. Fjeldly

Norwegian University of Science and Technology
Kjeller, Norway

Michael S. Shur

Rensselaer Polytechnic Institute
Troy, NY

IEEE PRESS

WILEY-INTERSCIENCE

A JOHN WILEY & SONS, INC., PUBLICATION

Published by John Wiley & Sons, Inc., Hoboken, New Jersey.
Published simultaneously in Canada.

For general information on our other products and services please contact our Customer Care Department within the U.S. at 877-762-2974, outside the U.S. at 317-572-3993 or fax 317-572-4002.

Wiley also publishes its books in a variety of electronic formats. Some content that appears in print, however, may not be available in electronic format.

Library of Congress Cataloging-in-Publication Data:

Lab on the Web: Running real electronics experiments via the Internet
    Edited by Tor A. Fjeldly and Michael S. Shur

ISBN 0-471-41375-5

Printed in the United States of America

10 9 8 7 6 5 4 3 2 1

CONTRIBUTORS

J. A. del Alamo, Massachusetts Institute of Technology, Cambridge, Massachusetts 02139

M. Billaud, Université Bordeuax I, 33405 Talence Cedex, France

L. Brooks, Massachusetts Institute of Technology, Cambridge, Massachusetts 02139

V. Chang, Massachusetts Institute of Technology, Cambridge, Massachusetts 02139

Y. Danto, Université Bordeuax I, 33405 Talence Cedex, France

H. Effinger, Fachhochshule Münster, 48565 Steinfurt, Germany

T. A. Fjeldly, Unik—University Graduate Center, Norwegian University of Science and Technology, N-2027 Kjeller, Norway

D. Geoffroy, Université Bordeuax I, 33405 Talence Cedex, France

F. Gomez, Universidad Autonoma de Madrid, 28049 Madrid, Spain

I. Gonzalez, Universidad Autonoma de Madrid, 28049 Madrid, Spain

I. Gustavsson, Blekinge Institute of Technology, S-372 25 Ronneby, Sweden

J. Hardison, Massachusetts Institute of Technology, Cambridge, Massachusetts 02139

L. Hui, Massachusetts Institute of Technology, Cambridge, Massachusetts 02139

F. Ingvarson, Chalmers University of Technology, S-412 96 Göteborg, Sweden

K. Jeppson, Chalmers University of Technology, S-412 96 Göteborg, Sweden

P. Lundgren, Chalmers University of Technology, S-412 96 Göteborg, Sweden

J. Martinez, Universidad Autonoma de Madrid, 28049 Madrid, Spain

C. McLean, Massachusetts Institute of Technology, Cambridge, Massachusetts 02139

G. Mishuris, Massachusetts Institute of Technology, Cambridge, Massachusetts 02139

T. A. Sæthre, Department of Physical Electronics, Norwegian University of Science and Technology, N-7491 Trondheim, Norway

W. Seifert, Fachhochshule Münster, 48565 Steinfurt, Germany

M. S. Shur, Electrical, Computer, and Systems Engineering and Physics, Applied Physics, and Astronomy, Rensselaer Polytechnic Institute, Troy, New York 12180

A. Skjelvan, Department of Physical Electronics, Norwegian University of Science and Technology, N-7491 Trondheim, Norway

A. Söderlund, Chalmers University of Technology, S-412 96 Göteborg, Sweden

A. Wiegand, Fachhochshule Münster, 48565 Steinfurt, Germany

C. Wulff, Department of Physical Electronics, Norwegian University of Science and Technology, N-7491 Trondheim, Norway

T. Ytterdal, Department of Physical Electronics, Norwegian University of Science and Technology, N-7491 Trondheim, Norway

T. Zimmer, Université Bordeuax I, 33405 Talence, Cedex France

CONTENTS

6   Remote Laboratory: Bringing Students Up Close to Semiconductor Devices   221

A. Söderlund, F. Ingvarson, P. Lundgren and K. Jeppson

Index   235

PREFACE

The ubiquity of the Internet as a communication medium has opened up a wide range of possibilities for extending its use into new areas. Remote education, a rapidly growing part of the university curricula, is one such area that can benefit greatly from the use of the Internet. Utilizing the Internet and the World Wide Web technology, courses can be offered to students anywhere in the world, without any other technical requirements than a personal computer and a telephone line. Although laboratory courses are an essential part of education, especially in engineering, such courses have until recently been considered impractical for remote students. Instead, these students often have to make time-consuming and expensive travel in order to complete laboratory courses. Also, with the tight budget at many educational institutions, lots of students are prevented from having local access to state-of-the-art equipment. However, the advances over the last decade in the Internet, World Wide Web technologies, and computer-controlled instrumentation presently allow net-based techniques to be utilized for setting up remote laboratory access, permitting remote education to be enhanced by experimental modules.

User-friendly, computer-controlled instrumentation and data analysis techniques are revolutionizing the way measurements are being made, allowing nearly instantaneous comparison between theoretical predictions, simulations, and actual experimental results. An ever-increasing array of industry-standard design and simulation tools now provides the opportunity to fully integrate the use of computers directly in the laboratory. Once this integration happens, it will no longer be crucial to have a piece of equipment physically located next to an engineer or scientist, thus opening the door for remote access via the Internet.

The remote, web-based experimentation augments the laboratory experience of the students by offering access to sophisticated instrumentation. It provides a natural and valuable extension of the traditional laboratory component, which normally uses relatively simple equipment. For limited periods of time, direct physical access to the Internet laboratory stations might be allowed in order to further acquaint the students with the equipment. However, remotely, the access might be 24 hours a day, and it might be further enhanced by live video showing the laboratory equipment and devices under the test. Remote laboratories have an enormous throughput, since, in many cases, the equipment is accessed by the user only for the very short time required for the actual measurement. All data processing occurs at the server and/or at the user's personal computer, so that he or she has the feeling of using the equipment alone. Another big advantage is safe and foolproof operation of expensive laboratory equipment, with safeguards built into the software.

These important advantages make remote-distance, interactive experimentation an important emerging educational trend. The Internet is an ideal medium for remote instruction purposes, offering interesting possibilities for disseminating many kinds of educational material to students, both locally and as part of remote education. Its ubiquity and protocol standards make data communication and front-end graphical user interfaces easy to implement.

Rensselaer Polytechnic Institute (RPI) and the Norwegian University of Science and Technology (NTNU) jointly developed a remote characterization laboratory for measurements of electronic devices over the Internet called AIM-Lab (Automated Internet Measurement Laboratory) in 1997. The time was apparently ripe for this development, since, at about the same time, several other groups independently started similar activities. Some of the remote laboratory systems and applications emerging from those pioneering activities are reviewed in this book. In fact, the independent nucleation of such activities has resulted in an interesting diversity of system solutions. In Chapter 1, the remote laboratory installations AIM-Lab (RPI) and LAB-on-WEB (NTNU) are presented, both resulting from the U.S.–Norway collaboration. Chapter 2 deals with WebLab developed at MIT Microelectronics. These two chapters deal with efficient systems for semiconductor device and circuit characterization. Chapter 3 describes the Retwine project, a collaboration between the University of Bordeaux I (France), the University of Applied Sciences of Münster (Germany), and the Autonomous University of Madrid (Spain). This project emphasizes training in the use of advanced instruments for controlling real remote experiments by means of virtual instruments with realistic images and functions. Chapter 4 presents the Next Generation Laboratory (NGL) at NTNU. Also emerging from the initial U.S.–Norway collaboration, NGL applies the .NET technology in remote characterizing of analog integrated circuits. The Remote Laboratory at the Blekinge Institute of Technology (Sweden), presented in Chapter 5, is applied to basic circuit characterization and to transducer experiments. Finally, the I-Lab system at Chalmers University of Technology (Sweden), which is discussed in Chapter 6, is used for the precise characterization of two-terminal devices. From the material presented in these chapters, the reader can compare different technologies involved in existing Web-based laboratories and review the rich variety of accompanying experiments.

The above remote laboratory installations are presently used to provide access to state-of-the-art laboratory instrumentation and experiments for both local and remote students. Presently, both national and international collaboration on remote laboratory development between universities is taking place, and international educational conferences have started to hold special sessions on this technology. This way, a broad range of sophisticated experiments is being made accessible to students on a global scale at a relatively modest investment for each institution. As remote laboratories have become operational at several sites, novel pedagogical uses have also emerged, including experimental demonstrations to enhance traditional classroom lectures, adding laboratory modules as homework exercises in regular courses, and establishing studio classrooms where students do supervised laboratory exercises on individual terminals. This remote experimentation greatly improves the learning process and encourages individual student discovery. All of this fits well into a strategy for distance learning.

No doubt, this technology should and will be applied to other areas of engineering and science, well beyond electrical circuits or device applications. Eventually, Internet laboratory courses covering many disciplines of engineering and science may be offered to students worldwide, removing a major obstacle for establishing a boundless and nearly complete remote education engineering curriculum and making engineering and science education attractive and available to segments of the population that otherwise would be disadvantaged by distance and lack of resources. This will be nothing less than a true revolution in distance education.

We hope that this book will be useful for students, teachers, and professors interested in remote instruction as well as for university and educational administrators who are interested in the development of efficient and economical educational technologies serving both their local student population and also students worldwide, including underprivileged communities. We would like to inspire teachers and professors to use the remote laboratory technology for applications in education and beyond, including many applications we have not even imagined.

TOR A. FJELDLY

Unik—University Graduate Center
Norwegian University of Science and Technology
N-2027 Kjeller, Norway

MICHAEL S. SHUR

Electrical, Computer and Systems Engineering and Physics,
Applied Physics and Astronomy
Rensselaer Polytechnic Institute
Troy, New York 12180-3590

ELECTRONICS LABORATORY EXPERIMENTS ACCESSIBLE VIA INTERNET

T. A. Fjeldly

Unik—University Graduate Center
Norwegian University of Science and Technology, N-2027 Kjeller, Norway

M. S. Shur

Rensselaer Polytechnic Institute, Troy, New York 12180

1.1 INTRODUCTION

Remote-distance, interactive learning is an important emerging educational trend. The Internet is an ideal medium for remote instruction purposes, offering interesting possibilities for disseminating educational material to students, both locally and as part of remote education. Its ubiquity and protocol standards make data communication and front-end graphical user interfaces relatively easy to implement.

Laboratory experiments are an indispensable part of engineering education that until recently have been considered impractical for distance learning. However, the advances over the last decade in the Internet, World Wide Web tech-

Lab on the Web: Running Real Electronics Experiments via the Internet
Edited by Tor A. Fjeldly and Michael S. Shur
ISBN 0-471-41375-5   Copyright © 2003 John Wiley & Sons, Inc.

nologies, and computer-controlled instrumentation presently allow net-based techniques to be utilized for setting up remote laboratory access, permitting remote education to be enhanced by experimental modules.

Currently, remote educational laboratories over the Internet, particularly in the area of electronics and instrumentation, have become operational at several sites (see Shen et al., 1999; Gustavsson, 2001; Geoffroy et al., 2001; Berntzen et al., 2001; del Alamo et al., 2002; Wulff et al., 2002; Söderlund and Jeppson, 2002). With these facilities, novel pedagogical uses have also emerged, including experimental demonstrations to enhance traditional classroom lectures, adding laboratory modules as homework exercises in regular courses, establishing studio classrooms where students do supervised laboratory exercises on individual terminals, and encouraging individual discovery activities among students. All of these activities fit nicely into a modern strategy for distance learning.

In a broader perspective, the Internet lab technology can be offered to remote students on a global scale, removing a major obstacle for establishing a boundless and complete remote engineering education curriculum. As an added benefit, such systems may offer students the opportunity to work with sophisticated equipment of the kind they are more likely to find in an industrial setting and that may be too expensive for most schools to purchase and maintain for educational purposes.

Our work on remote lab systems started in 1997 as a collaboration between Rensselaer Polytechnic Institute (RPI) in Troy, New York, and the Norwegian University of Science and Technology (NTNU) in Norway. Presently, we operate three sites: AIM-Lab (Automatic Internet Measurement Laboratory) at RPI, LAB-on-WEB at UniK—University Graduate Center near Oslo, Norway (affiliated with NTNU and the University of Oslo), and NGL (Next Generation Laboratory) at NTNU in Trondheim, Norway. At these sites, we have explored and developed different system technologies, as will be explained below and in Chapter 4 of this book. However, all systems are based on a server–client architecture, where the clients (students) communicate over the net with a server and its experimental setup using modern web browsers. For the most part, we have developed dedicated system software that does not require any download by the client, but in some cases optional solutions are offered based on software that can be downloaded from the Internet for free.

Biased by our background in physics and electronics, we have emphasized the establishment of laboratories dedicated to semiconductor device characterization, with experiments performed on microelectronic test chips and on commercial devices. Our labs have been used in courses on semiconductor devices and circuits at the senior or first-year graduate level at all our institutions. At our sites in Norway and the United States, we have jointly investigated several practical solutions for establishing such laboratories. Central objectives were to create a user-friendly and efficient technology for interactive, on-line operation of the lab experiments, to allow communication with minimum overhead, to provide a functional client interface, to establish a variety of experiments, and to allow flexibility in configuring experiments from the client side.

The AIM-Lab site at RPI is based on a TCP/IP (Transmission Control Protocol/Internet Protocol) communication solution, which uses a Java applet on the client side. This was achieved by means of a JVM (Java virtual machine) in the web browser that can download and execute Java code. The client sees a pop-up window that provides interaction and communication directly with the server. This system is described in Section 1.2.

LAB-on-WEB at the UniK site relies on modern web and instrument control technologies, including COM+ (component object model with extensions), ASPs (active server pages), ISAPI (Internet server application programming interface), and LabVIEW (Laboratory Virtual Instrument Engineering Workbench) version 6i from National Instruments. Advanced functionalities of modern web browsers are utilized, allowing the system to communicate in terms of XML (eXtensible Markup Language) and SVG (scalable vector graphics). SVG is a vector-based, open-standard file format developed by the World Wide Web Consortium, which represents a new generation of dynamic, data-driven, and interactive graphics. The LAB-on-WEB system is described in Section 1.3.

At the NGL site at NTNU, Microsoft's new .NET technology was adopted. This solution is described in Chapter 4 of this book.

In Section 1.4, we discuss by way of examples some of the experiments available at AIM-Lab and LAB-on-WEB.

This presentation emphasizes the technological aspects of our remote lab systems and the specific experiments offered. We have so far not performed systematic evaluations of the educational benefits derived from using these labs. Obvious benefits to the students are apparent in terms of availability and ubiquity, likewise the presentation of "live" lab demonstrations in the classroom. Our subjective impressions are that the students show a positive interest and curiosity, and we register a positive attitude to the freedom offered in terms of scheduling their lab sessions.

1.2   AIM-LAB AT RPI

1.2.1   System Architecture

The AIM-Lab system architecture is shown in Figure 1.1. The server, written in Microsoft Visual C++, includes two main components. One of them is a TCP/IP server socket that receives commands sent over the Internet. The second component, the driver interface layer (DIL), interfaces between the instrument driver and the higher levels of the server (Shen et al., 1999; Fjeldty et al., 2000). The DIL sends the commands to the instrument driver, which uses the GPIB (general-purpose instrument bus) Institute of Electrical and Electronics Engineers (IEEE) 488.2 standard protocol to drive the instruments. A third com-

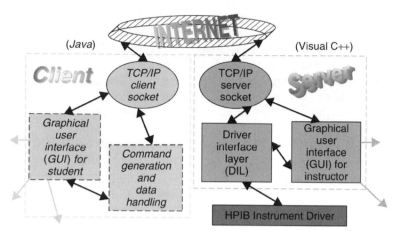

Figure 1.1   System configuration for AIM-Lab using TCP/IP communication (after Shen et al., 1999).

ponent is the GUI (graphical user interface) for the instructor. This interface on the server side allows the instructor to monitor and control the server process as well as modify the configuration of the instrumentation.

The client side is a Java applet that initially appears as a push button on the AIM-Lab web page. By pressing the button, the applet creates a pop-up window that provides the GUI interface to the user. The client's command generator issues commands according to the parameter set specified by the user and sends them via the TCP/IP client socket to the server. The experimental results sent back by the server are then handled and displayed in the client window.

In AIM-Lab, Java is the programming language of choice on the client side, since it offers the flexibility of a GUI design, convenient network programming, and platform independence. The operation is achieved by means of a JVM in the client's web browser that can download and execute Java code. The client sees a pop-up window that, on the one hand, provides GUI interactions to the user and, on the other hand, communicates directly with the server. The GUI interface is created according to the information on the experiments received from the server upon client initialization.

The AIM-Lab system is designed to minimize the overhead of the data communication through the Internet, maximize the server performance and efficiency, ensure the data accuracy and integrity, and provide an easy access to the user. In order to maximize the server performance and efficiency, we developed the server as a Windows-based MDI (multidocument interface) application. This is a multiuser and multiexperiment environment with a task queue. For each user, it records all the commands and data in a dedicated document window.

The experiment requests are sent to the instrument driver in the order of receipt, and the resulting data are sent back accordingly. No experiment failure or error caused by the clients leads to a malfunction of the server. Any experiment that takes an exceedingly long time to finish (which might suggest a failure) is discarded and hence does not affect the other experiments. The server does not parse or interpret the commands. It assumes that the command generator of the client program correctly generates the commands. In case of an error, the server will discard the commands as described above. This reduces the processing overhead of the server and makes the server very flexible. When an instrument or a circuit is changed, the server sends the information about the change to all the running clients.

1.2.2   AIM-Lab Operation

The system provides easy access for the user and maximizes the speed of the on-line measurements. No file needs to be downloaded in order to perform the experiments, which is an advantage compared to other realizations in terms of speed and security. All the user has to do is to access the AIM-Lab website (`http://nina.ecse.rpi.edu/shur/remote/`) and start the client window by pressing a button. The user can set up experiments and send experiment requests by activating pop-up dialog boxes. The results of the measurements are displayed in the client window, and the user can navigate between the experimental plots with ease. The resulting data and plots can be saved using `Copy` and `Paste` functions of the Windows and Unix systems.

The communication overhead is minimized by sending only the absolutely necessary information via the Internet and by organizing the generated commands, data results, and server messages in the most compact format. We have tested the system off-campus using a commercial 56-kbps modem. According to our test, the time needed to access the website and start the client window is about 50 sec. It takes less than 10 sec for the system to perform a complete experiment, including sending commands and receiving and plotting the data.

The user accesses the AIM-Lab website using a standard web browser. When logging in, the server application is launched and all information on the available experiments is read from a library file and sent to the client. The client window behaves as a stand-alone application in which the user selects an experiment and specifies experimental parameters (e.g., voltage range, step size) in pop-up dialog boxes. Consecutive experiments can be set up and sent to the server without waiting for the previous experiment to finish. An experiment is initiated by activating the `Start Experiment` menu item in the `Operation` pop-up menu. Some of the client windows encountered are shown in Figure 1.2.

The instructions from the client are then sent via the TCP/IP client socket to the server, which runs the experiment and returns the experimental data to the client. The results are presented in the client browser window as columns of numerical data and a graph. The numerical data can be copied from the window for further processing by the user. As an example, Figure 1.3

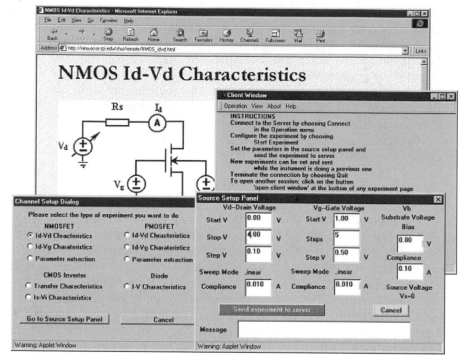

Figure 1.2 AIM-Lab client windows with information on experiment (background), with menus and instructions for running experiment (middle right), for selecting experiment (lower left), and with panel for setting experimental parameters and initiating experiment (lower right).

shows the current–voltage $(I-V)$ characteristics of an n-channel metal–oxide–semiconductor field effect transistor (MOSFET).

Java applets provide good control. However, unsigned applets make it awkward for the client to store and present data received from the server side and to transfer them to other applications (except by cut-and-paste) because of Java's security structure. A further problem with Java is that the functionality of an applet may vary between different browsers. While Java 2 has better support for the user interface, some of the new classes included in this version are not automatically compatible with JVM, requiring an additional plug-in to be installed in the browser. For many potential users, this is a problem, partly because of skepticism toward plug-ins and partly because of local security regulations in many organizations. Besides, the future support of Java is uncertain.

1.2.3 AIM-Lab Experimental Setup

1.2.3.1 Device Test Structures AIM-Lab is presently dedicated to the characterization of a group of test devices that includes a set of complementary metal–oxide–semiconductor (CMOS) devices and a set of light-emitting diodes

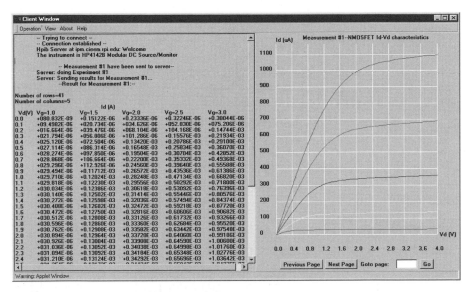

Figure 1.3 Example of AIM-Lab client-side window with experimental results (NMOS *I–V* characteristics) corresponding to instructions submitted by client (see Figure 1.2).

(LEDs) made from different compound semiconductors. CMOS is the most important integrated circuit technology, far outselling all other semiconductor technologies, including bipolar, thin-film transistor (TFT), and compound semiconductor technologies. The importance and proliferation of CMOS necessitate a good understanding of its operation by very large scale integrated (VLSI) designers and users alike. The best way to teach the basics of CMOS technology is by a hands-on approach, which combines the basic theory of operation with measurements, parameter extraction, and CMOS circuit simulation (see Lee et al., 1993; Fjeldly et al., 1998).

AIM-Lab presently allows experiments to be performed on the CMOS test chip shown in Figure 1.4. This chip, designed, fabricated, and characterized by our group, includes two arrays of NMOS (*n*-channel MOSFET) and PMOS (*p*-channel MOSFET) devices with a wide range of gate geometries for the purpose of investigating the scaling properties of the devices. In each array, all source electrodes are interconnected, so are all the gates and all the substrate electrodes. Only the drain electrodes have separate contact pads. Additional diagnostic structures are also available, including a loaded and an unloaded ring oscillator. Since each experimental configuration is hard-wired at the server site, only a limited number of the possible experiments are available to the remote user at any given time. Presently, six experiments are specified, involving measurements of MOSFET and inverter characteristics used for device characterization.

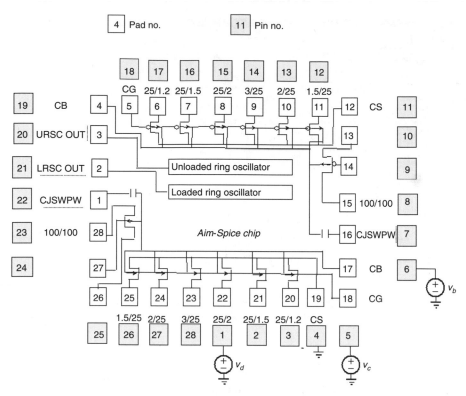

Figure 1.4 Layout of AIM-Spice CMOS test chip. PMOS and NMOS transistors of different dimensions are shown in upper and lower sections of chip, respectively. Gate width W and length L in micrometers are shown as W/L next to each MOSFET. Biases shown are those used for obtaining current–voltage characteristics of NMOS with $W/L = \frac{25}{2}$. (Shen et al., 1999).

Additional AIM-Lab experiments involve LEDs that emit light of different colors—green, yellow, and blue. The colors, or wavelengths, of the emitted light are related to the electronic band gaps of the compound semiconductor materials from which the diodes are fabricated. The band gaps can be roughly estimated from the turn-on voltages of the diodes, as observed by measuring the I–V characteristics. The emitted light colors from the different LEDs can be observed with a web camera.

1.2.3.2 AIM-Lab Instrumentation The experimental instrumentation in AIM-Lab consists of a Hewlett-Packard (HP 4142B) direct-current (DC) source/monitor with one source monitor unit (SMU) for each separate voltage source or measurement node. The sever is an HP D3695 host computer using the Microsoft NT operating system with Peer Web Services installed. An GPIB instrument driver installed in the host personal computer (PC) handles the communication with the HP 4142B instrument. The HP 4142B has a very high resolution and speed and an embedded processor that accepts high-level commands. However, this equipment is expensive and not very portable. In an alternative configuration, we have also investigated the use of peripheral component interconnect/industry standard architecture (PCI/ISA) cards that can be installed in the expansion slots of the host PC (see Shen et al., 2000). In this case, the HPIB driver and the HP 4142B instrument were replaced by low-cost voltage output and data collection cards. These cards have simple functionalities for outputting a single voltage and converting it into a digital signal that can be read by the host PC. However, they suffer from lower resolution. Besides, the sampling resistors for the current measurements have to be selected with care.

1.3 LAB-ON-WEB AT UNIK

LAB-on-WEB at UniK in Norway was developed in several phases from 1999 to the present, mostly through student master's theses and externally funded projects, notably by Nordunet2, a program financed by the Nordic Council of Ministers. We initially explored alternative network and server solutions to those used in AIM-Lab (see Section 1.2), considering both ActiveX-based solutions, such as ISAPI and COM+, and LabVIEW solutions based on the Internet-adapted version 6i. Later, we investigated ways to use switching matrices for the purpose of expanding the number of available experiments and to enable on-line reconfiguration of simple circuits by the clients. The objective of the latter is to provide the students with more freedom to pursue "individual discovery."

The first complete lab system was built around the CMOS test chip shown in Figure 1.4. The central measurement instrument was the HP 4142B modular DC source/monitor with HP 41421B source monitor units (SMUs). Presently, we have replaced the test object by a more comprehensive CMOS lab-on-a-chip unit that allows a much richer variety of experiments to be performed (see Section 1.3.1.1). To fully utilize this potential, new instruments have also been added to the LAB-on-WEB setup.

1.3.1 LAB-on-WEB Experimental Setup

1.3.1.1 Alfa Chip Test Structure The present version of LAB-on-WEB includes a comprehensive electronic device laboratory on a chip that significantly enhances the learning experience of our remote lab. This so-called Alfa chip was developed as part of project number 3.0284.5 within the ALFA exchange program between universities in the European Union and Latin America under the auspices of the European Commission (see Delmas Bendhia et al., 1998). The project had participation from France (E. Sicard), Germany (F. Schwiertz), Mexico (E. Gutierrez), Norway (T. A. Fjeldly), Spain (E. G. Moreno), and Venezuela (A. O. Conde). The integrated circuit was designed by the ALFA partners using the Microwind* tool developed by Institut National des Sciences

---

*Microwind is a PC freeware tool featuring layout design and built-in analog simulation.

Appliquée (INSA), Toulouse (Sicard, 2000). The chip was fabricated at ATMEL Rousset in France using a two-metal 0.7-μm CMOS process.

The Alfa chip contains diodes, capacitors, transistors, inverters, and other test structures, all fabricated in CMOS technology. Besides providing a valuable familiarity with the static and dynamic behavior of the building blocks of modern integrated circuits, the lab is also designed for the extraction of device and processing parameters needed for circuit design. The extracted parameters are suitable for use with circuit simulators such as AIM-Spice (`http://www.aimspice.com`; see also Fjeldly et al., 1998) to predict the behavior of more complicated circuits such as the ring oscillator also included on the chip.

The Alfa chip layout is shown in Figure 1.5, where the relevant devices used by LAB-on-WEB are highlighted. The chip is packaged in a pin gate array

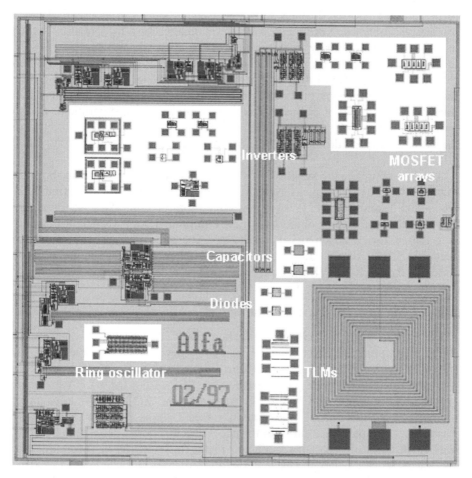

Figure 1.5   Alfa chip layout. Devices in highlighted areas are used in LAB-on-WEB (Fjeldly et al., 2002).

NMOS

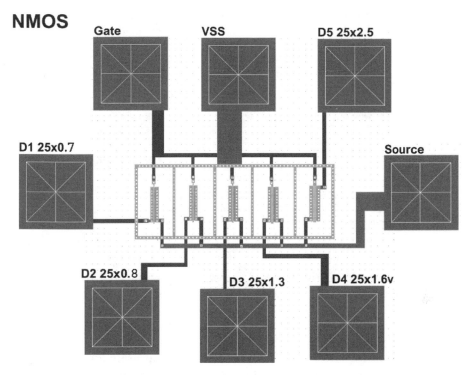

Figure 1.6 Layout of NMOS array. All gates are 25 μm wide while gate lengths vary between 0.7 and 2.5 μm. The Alfa chip also contains a PMOS array with the same geometries (Fjeldly et al., 2002).

(PGA) 144 chip carrier and the various contact pads, each of size $40 \times 40 \ \mu m^2$, are bonded to the carrier pins.

Figure 1.6 shows a more detailed view of the array of NMOS transistors located in the upper right corner of the Alfa chip. It contains five n-channel MOSFETs with a gate width W of 25 μm and gate lengths L varying from 0.7 to 2.5 μm. The MOSFETs have separate drain contacts, but all the gate electrodes are interconnected and so are the source contacts. The Alfa chip also contains a PMOS array with the same geometries. These arrays are suitable for characterization of individual MOSFETs, for extracting their model parameters, and for studying the effects of geometric scaling of these devices.

The Alfa chip includes two separate CMOS inverters with different geometries (upper left area of Figure 1.5). The one shown in Figure 1.7 has $W = 25$ μm and $L = 0.8$ μm for both the NMOS and the PMOS transistor. The second inverter has the same gate length but a gate width of 1.2 μm. Note that inverters with other geometries can also be formed by making suitable external connections between devices in the two MOSFET arrays discussed above.

A ring oscillator consists of a cascade of an odd number of inverters, where the signal from the last inverter is fed back into the first one. The individual

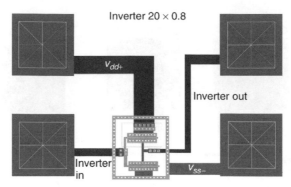

Figure 1.7 Layout of CMOS inverter with NMOS (lower) and PMOS (upper) devices both with gate width $W = 20$ μm and gate length $L = 0.8$ μm. Alfa chip also contains CMOS inverter with $W = 1.2$ μm and $L = 0.8$ μm (Fjeldly et al., 2002).

elements are visible in Figure 1.8. The oscillation is triggered by an external signal input at the enable pad and is sensed at the output pad. The oscillation frequency is the inverse of the sum of the signal delays experienced by each inverter. The NMOS gate geometry is $W = 1.6$ μm and $L = 0.8$ μm, and the PMOS gate geometry is $W = 2.4$ μm and $L = 0.8$ μm. The difference in gate widths between the NMOS and PMOS serves to compensate for differences in electrical properties of the two types of devices. Hence, from a measurement of the oscillation frequency and using the number of inverters in the device, the single-stage delay can be extracted. This is a very important parameter that expresses the speed of digital signal transmission between the elements.

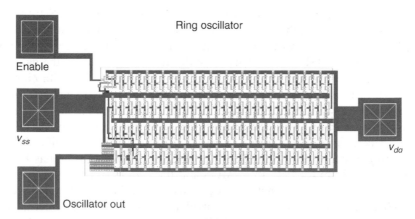

Figure 1.8 Layout of CMOS ring oscillator consisting of 110 inverters and one NAND gate at Enable pad. NMOS gate geometry is $W = 1.6$ μm and $L = 0.8$ μm, and PMOS gate geometry is $W = 2.4$ μm and $L = 0.8$ μm (Fjeldly et al., 2002).

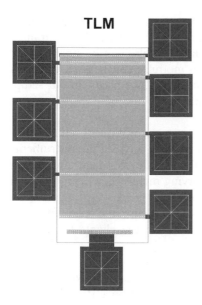

TLM

Figure 1.9 TLM pattern for determining contact and sheet resistance of ohmic contact regions (Fjeldly et al., 2002).

The structure shown in Figure 1.9 consists of a series of unevenly spaced electrodes across a region of doped material, typically of the same kind as that used in the ohmic source and drain regions of the MOSFETs. This structure is called a transmission line model (TLM) (see Shur, 1995). By measuring the total resistances between the adjacent contacts and plotting these versus their contact separations, the contact resistance can be determined by extrapolating the resulting straight line to zero separation. From the slope of this plot, the sheet resistance of the doped semiconductor region is found. The TLM regions are 120 µm wide and the distances between adjacent contacts are 5, 15, 25, 35, 45, and 45 µm.

In addition, the Alfa chip contains a pair of capacitors and a pair of diodes, which are very useful for the extraction of important parameters such as gate oxide thickness, parasitic capacitances associated with source and drain, flat-band voltage, and substrate doping.

Note that not all the devices described above will be made available at LAB-on-WEB at any given time. At times, LAB-on-WEB also will include experiments on other semiconductor devices and structures, including bipolar devices such as diodes and junction transistors.

1.3.1.2 Instrumentation The LAB-on-WEB server is a Dell Power Edge 4300 computer with a 500-MHz Pentium III processor, a Windows 2000 server operating system, and an MIIS (Microsoft Internet Information Server) version 5.0. The server and the instruments are connected via a GPIB.

The main instrument is an HP 4142B modular DC source/monitor, which is the same type of instrument as in the AIM-Lab setup. This is a high-speed, accurate, and computer-controlled DC parametric measurement instrument for characterizing semiconductor devices. Voltages and currents can be applied or measured within 4 msec, and up to 1023 data samples can be stored in the internal memory. Up to eight different plug-in modules can be used with this instrument, allowing us to tailor the instrument to suit our needs. In our setup, three HP 41421B SMUs are installed, in addition to the built-in 0-V source GNDU (ground unit).

By means of a switch matrix, the client is given the option to remotely re-configure the experiment by changing the connections between the instruments and the CMOS test chip. This technique is still under development, and only a limited range of configurations are possible, serving merely as a demonstration of the functionality and potential of this approach.

A Tektronix TDS 3052, 500-MHz oscilloscope is also included, for both servicing and testing the setup but also for allowing the clients to measure transient events or waveforms.

Experimental Setup Version 1   The first version of the LAB-on-WEB server/ laboratory setup is shown in Figure 1.10. Here, we used an HP 34970A data

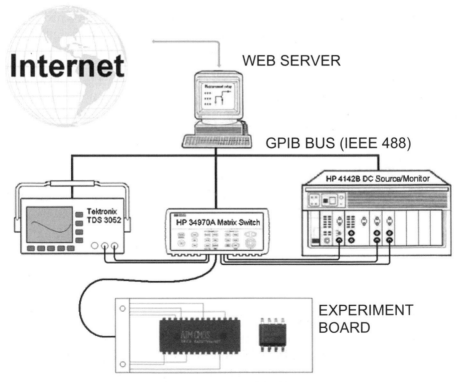

Figure 1.10   Setup of LAB-on-WEB version 1. Test objects to left on experiment board is AIM-Spice CMOS test chip (see Figure 1.4).

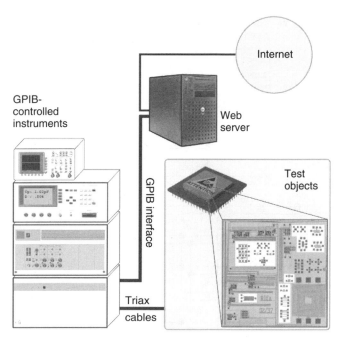

Figure 1.11 Setup of LAB-on-WEB version 2. Instruments to left are, from below, Agilent E5250A low-leakage switch unit, HP 4142B modular DC source/monitor with HP 41421B SMUs, Agilent 4284A precision *LCR* meter, and Tektronix TDS 3052, 500-MHz digital oscilloscope. Test object is Alfa chip discussed above (Fjeldly et al., 2002).

acquisition/switch unit with HP 34904A plug-in modules as a switch matrix. However, this instrument adds a fair amount of noise to the measurements, making it difficult to measure current levels below about 10^{-5} A. This is unacceptable for a precise characterization of reverse-bias currents in bipolar devices and of subthreshold behavior in field effect transistors (FETs). An alternative to HP 34970A is the much more expensive instrument Agilent E5250A low-leakage switch unit used in the second version of LAB-on-WEB (see below).

Experimental Setup Version 2   The current experimental setup of the LAB-on-WEB server/laboratory unit is shown in Figure 1.11. It includes the same instruments as in version 1, except that the HP 34970A data acquisition/switch unit was replaced by the Agilent E5250A low-leakage switch unit to allow measurements of ultralow currents (the former instrument can, of course, still be used in parallel with the new one for less sensitive applications). The switch mainframe is equipped with two E5252A 10×12 matrix switch plug-in modules, for a total of 10 inputs and 24 outputs. A maximum of four modules can be installed, creating a 10×48 matrix. Figure 1.12 shows a block diagram of

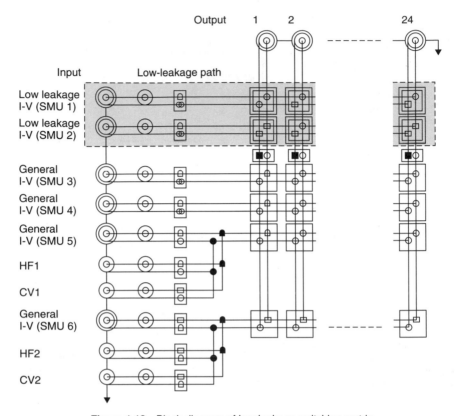

Figure 1.12   Block diagram of low-leakage switching matrix.

the low-leakage switching matrix. Among the 10 inputs are 2 each for low-noise $I–V$, capacitance–voltage $(C–V)$, and high-frequency measurements.

The present setup also includes an Agilent 4284A precision LCR (induction, capacitance, and resistance) meter for measuring $C–V$ characteristics.

1.3.2   LAB-on-WEB Architectures

Compared to AIM-Lab, LAB-on-WEB is based on more flexible solutions that utilize the rich functionalities of modern web browsers, allowing the server system to respond in many different formats, such as JavaScript, HTML (HyperText Markup Language), XML, and SVG, which give the client great flexibility in storing, processing, and presenting the data received (Berntzen et al., 2001). Among the server-side web solutions investigated are ISAPI server extensions COM+ (Strandman et al., 2002) and .NET (see Chapter 4 of this book).

Another alternative investigated is the use of the LabVIEW 6i software from National Instruments (Berntzen et al., 2001). In this solution, the server runs a full version of LabVIEW 6i, which incorporates Internet communication capabilities and functionalities to access and control instruments and to obtain and return data. The client can communicate with the server and the experimental setup in two ways: by means of a web browser, which runs a dedicated CGI (common gateway interface) script in the server, or using the LabVIEW Player (alternatively, the full version of LabVIEW 6i).

1.3.2.1 *ISAPI Server Extensions*

The ISAPI server extensions enhance the capabilities of the HTTP (Hypertext Transport Protocol) server included in the MIIS, allowing browser programs to interact with scripts or separate executable programs running on the server. Compared to the CGI standard incorporated into HTTP, the ISAPI extensions are faster and have added functionality.

ISAPI server extensions are implemented as a DDL (dynamic link library) file on the web server. For our purpose, they are programmed in Visual C++ by means of a wizard that comes with Microsoft Visual Studio. In response to calls from the client, they return HTML or XML code that can include JavaScripts or applets, allowing the data to be presented as rich graphs and tables in the client terminal.

In addition to the ISAPI extensions, different kinds of ISAPI filter functions can be implemented to perform various useful tasks. Port management and encryption filters can provide access control, data encryption, and server authentication. The log filter can trap client information, from which an activity log can be created. The page translation filter can be used to transform XML to, for example, HTML or WML (Wireless Markup Language) using XSL (Extensible Stylesheet Language) stylesheets. The XSL transformation can take place either on the client side or in the server based on logged information about the client web browser.

The ISAPI solution is shown schematically in Figure 1.13. This solution was tested and found to be quite satisfactory. However, it is presently not implemented in LAB-on-WEB since the solutions discussed below were found to be superior in several respects.

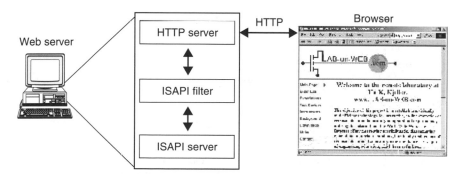

Figure 1.13 ISAPI architecture (Strandman et al., 2002).

1.3.2.2 COM+ Solution COM+ is the second generation of COM (component object model), which is a set of services that allows us to create object-oriented, customizable, and upgradeable, component-based, distributed applications. One of the primary goals of COM is to ease the creation of multitier applications that can be reused from any programming language. COM+ includes added features that make COM easier to use and simplifies the development. Some of these features are component load balancing, just-in-time activation, asynchronous method invocation, in-memory database, queued components, and improved administrative services.

COM+ applications are typically middle-tier applications; that is, they move information between clients and back-end resources such as databases (in our case, measurement data from instruments). The COM+ component developed for the remote laboratory is a library application that is compiled to a DLL and is loaded into the process of the client. The LAB-on-WEB communication process is indicated in Figure 1.14. In this architecture, the COM+ component

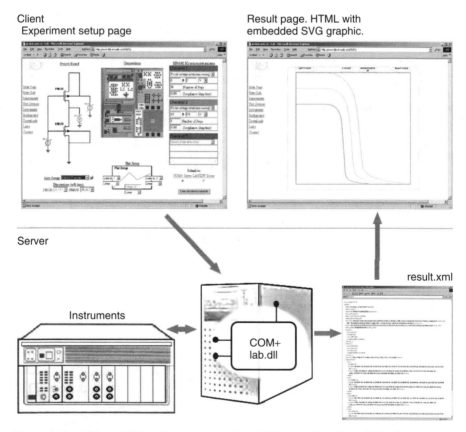

Figure 1.14 LAB-on-WEB architecture based on COM+ solution. From SVG client interface (upper left), client calls COM+ component via ASP. Measurement data are returned in XML format and presented in client browser (upper right).

is called from an HTML form in the active client page (see below) via an ASP, initiating the transfer of specifications on the desired measurement series to be performed. Subsequently, the instrument receives the necessary commands to perform the measurements. When the series is completed, the ASP receives the measurement results from the component and relays them to the client in XML format. The presentation of the results in the client browser can, for example, be in the form of an SVG plot along with a table of the raw data. The graph shown in the upper right of Figure 1.14 is an example of how data can be presented in the client browser using the SVG graphics.

LabVIEW version 6i from National Instruments is a powerful graphical programming development environment for data acquisition and control, data analysis, and data presentation. It comes with advanced Internet-ready capabilities and the concept of measurement intelligence, which includes automatic measurement hardware configuration for fast application development. Instead of writing program code, we build virtual instruments (VIs) with front-panel user interfaces that may contain, for example, numeric displays, meters, charts, and advanced graphs. The functionality is specified in block diagrams. The VIs allow control of any GPIB instrument.

In LAB-on-WEB, the client may execute experiments in two different modes, either with the executable LabVIEW Player (alternatively with the full LabVIEW 6i) or using a web browser.

LabVIEW Player is available free of charge from National Instruments. It is designed specifically for sharing measurement and automation knowledge across the Web in the form of secure LabVIEW VIs. By downloading and installing LabVIEW Player, the client can open and run LabVIEW VIs located in the LAB-on-WEB server. Player is equipped with functionalities for processing the data received to fit the user's need. Hence, clients obtain test results and measurement data from remote locations by means of built-in Internet tools. LabVIEW Player is richly equipped with functionalities for manipulating the data received and the graphs generated to fit the user's need. When further analysis is needed, the experimental data received by the client can be forwarded to an Excel document with the push of a button.

A drawback of LabVIEW Player is its considerable size, about 17.2 MB. This makes it inconvenient for downloading via a telephone modem for home use.

Using the LabVIEW Internet Developers Toolkit, it is possible to utilize the built-in HTTP server to make the VI front panels viewable from regular web browsers by means of CGI scripts. This solution provides additional flexibility in the transmission of measurement data, permitting different formats such as JavaScript, HTML, or XML. Hence, in this case, there is no need to download the large LabVIEW Player executable, allowing, for example, the interactive SVG client windows, such as the one shown in the upper right of Figure 1.14, to be used for configuring experimental setups. Figure 1.15 shows an example of a LabVIEW VI diagram with building blocks constituting a CGI solution.

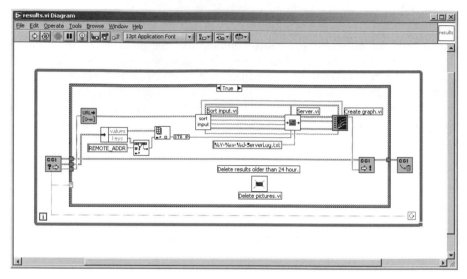

Figure 1.15   Block diagram of LabVIEW virtual instrument for CGI solution.

1.3.3   LAB-on-WEB Client Interfaces

An important aspect of using the Internet for educational purposes is to provide a user-friendly client-side interface designed to enhance the learning experience. In our case, we had to weigh functionality against the use of lablike impressions. However, we decided to use images of circuits and instrument to visualize the selections made by the client, including the interactive reconfiguration of the experiments. Two versions of the main LAB-on-WEB client interface, one used with the AIM-Spice CMOS chip and the other with the Alfa chip, are shown below. These interfaces were realized by means of JavaScript and the SVG file format and are designed for use with all the system architectures described above when the client uses a web browser. However, to achieve a full functionality, the SVG plug-in is required.

As discussed above (see Section 1.3.2.3), we have also implemented two additional client interfaces based on LabVIEW.

1.3.3.1   SVG Client Interface Version 1   Figure 1.16 shows the client interface for the AIM-Spice CMOS chip. Several experimental configurations are prepared for direct selection, as indicated in the table at the lower left-hand side. By clicking on one of the entries in this table, an arrow will be displayed to confirm the selection made, and the corresponding circuit diagram pops up in the middle of the page. However, by clicking on the various elements in the circuit diagram, a given circuit configuration may be further customized, and the external connections between the instrument and the test chip, shown as color-coded lines in the upper right of Figure 1.16, will be updated. The experimental parameter settings are entered in the tables in the lower right. The

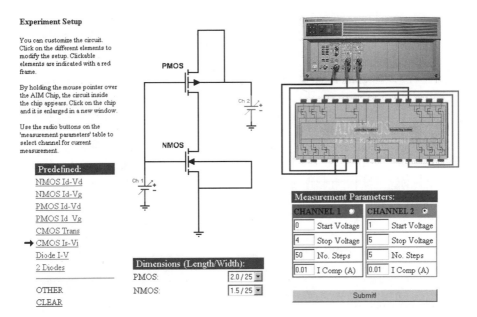

Experiment Setup

You can customize the circuit. Click on the different elements to modify the setup. Clickable elements are indicated with a red frame.

By holding the mouse pointer over the AIM Chip, the circuit inside the chip appears. Click on the chip and it is enlarged in a new window.

Use the radio buttons on the 'measurement parameters' table to select channel for current measurement.

Figure 1.16 Version 1 of interactive SVG client window used for configuring experiments based on AIM-Spice CMOS chip and for setting experimental parameters. Selections are submitted by pressing Submit panel.

external connections are physically implemented via the matrix switch once the Submit button is activated.

1.3.3.2   SVG Client Interface Version 2   The client interface for the Alfa chip is shown in Figure 1.17. In this case, an even wider range of predefined experimental configurations are prepared for direct selection from the drop-down table at the lower left of the page. When a selection is made, the corresponding circuit diagram immediately pops up in the field above, and the upper middle field zooms in on the relevant part of the chip. Here, the external connections are indicated by matching color coding of the device pads and the force/sense elements of the circuit diagram. Again, by clicking on the various elements in the circuit diagram, the configuration may be further customized. The experimental parameter settings are entered in the tables to the right.

Using Plot Setup in the lower middle field, the client can configure the presentation of the experimental results. The diagram can be selected with one or two vertical axes, which can be individually formatted in a linear or a logarithmic scale.

The experiment can be directed to either the COM+ server or the LabVIEW server using the buttons in the lower right of the client page. The external connections are physically implemented via the matrix switch once the Execute Measurement field is pressed.

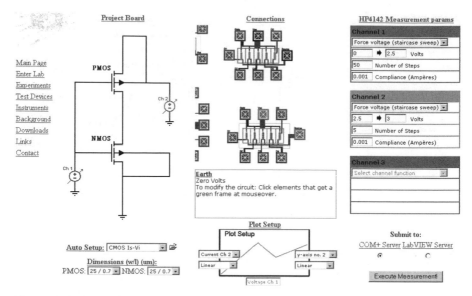

Figure 1.17 Interactive client interface for selecting and configuring experiments, setting experimental parameters, and specifying output graph. Selections are submitted by pressing `Execute Measurement` panel.

As an example of the SVG output using this client interface, Figure 1.18 shows the short-circuit current profiles obtained from an inverter connection in the Alfa chip.

1.3.3.3 LabVIEW Player Interface Figure 1.19 shows a LabVIEW VI client panel with a list of preselected experiments located in the upper left-hand side. Once an experiment is selected from this list, a corresponding circuit diagram pops up on the right. The experimental parameters are entered in the tables at the lower left side. Figure 1.20 shows an example of experimental CMOS inverter transfer characteristics obtained using the LabVIEW Player solution.

1.3.3.4 LabVIEW CGI Interface A dedicated client window for the LabVIEW CGI solution is shown in Figure 1.21. This interface can be selected from a web browser as an alternative to the LabVIEW Player interface of Figure 1.19. Figure 1.22 shows a presentation of NMOS characteristics obtained using the CGI interface.

By means of the built-in HTTP server, it is possible for the VI to respond to multiple clients and to continuously update their displays. This is an example of how the Toolkit libraries can be programmed to convert VIs into image files for

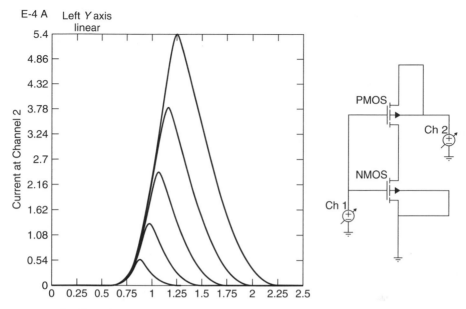

Figure 1.18 Short-circuit current profiles of inverter configuration of Alfa chip measured in Ch. 2 as function of input voltage at Ch. 1 at different supply voltages applied to Ch. 2.

display within HTML pages. Security levels can also be incorporated into the server to limit access to front panels and data. Likewise, access to the VIs can be password controlled based on the user's IP address.

1.4 EXAMPLE EXPERIMENTS

The following example experiments can be performed using either AIM-Lab located at RPI in Troy, New York, or LAB-on-WEB located at UniK near Oslo, Norway. The two sites can be accessed at http://nina.ecse.rpi. edu/shur/remote and http://www.lab-on-web.com/, respectively. Note that both system work best with MS Explorer.

The main page of AIM-Lab is shown in Figure 1.23. Using the various links on this page, information on the laboratory and the experiments can be obtained. The experiments can be accessed by selecting Connect to the measurement setup in the panel shown at the bottom.

The main page of LAB-on-WEB is shown in Figure 1.24. Using the various links on this page, information on the laboratory, the experiments, the instruments, and the test structures can be obtained. The client interfaces can be accessed by selecting Enter Lab in the panel to the left.

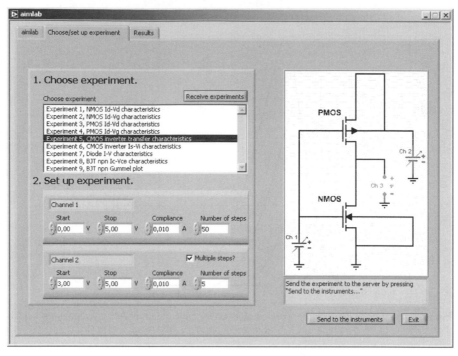

Figure 1.19 Interactive client interface for selecting and configuring experiments, setting experimental parameters, and specifying output graph. Selections are submitted by pressing `Send to Instruments` panel.

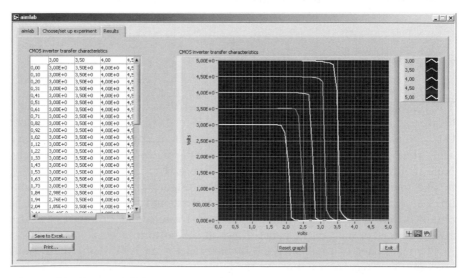

Figure 1.20 Measured CMOS transfer characteristics obtained using LabVIEW Player virtual instrument client window of Figure 1.19.

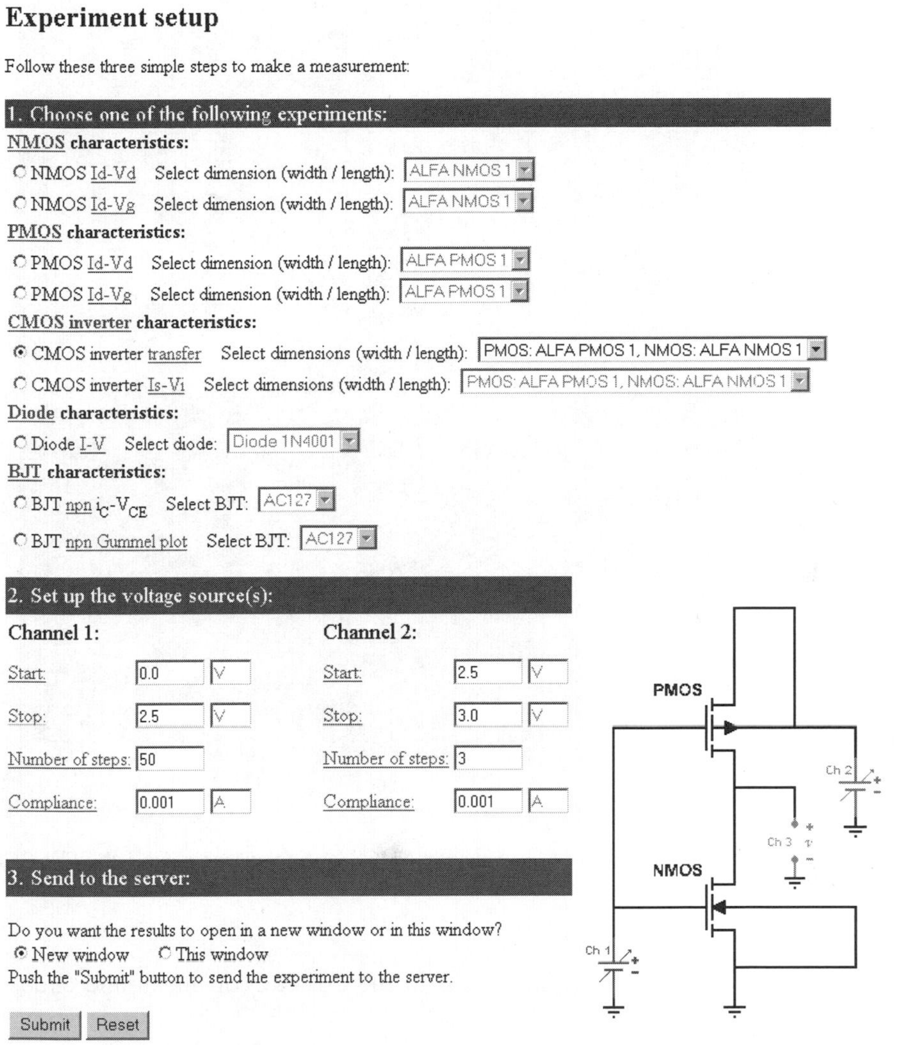

Experiment setup

Follow these three simple steps to make a measurement:

1. Choose one of the following experiments:

NMOS **characteristics:**

    ○ NMOS Id-Vd   Select dimension (width / length): [ALFA NMOS 1 ▾]

    ○ NMOS Id-Vg   Select dimension (width / length): [ALFA NMOS 1 ▾]

PMOS **characteristics:**

    ○ PMOS Id-Vd   Select dimension (width / length): [ALFA PMOS 1 ▾]

    ○ PMOS Id-Vg   Select dimension (width / length): [ALFA PMOS 1 ▾]

CMOS inverter **characteristics:**

    ◉ CMOS inverter transfer   Select dimensions (width / length): [PMOS: ALFA PMOS 1, NMOS: ALFA NMOS 1 ▾]

    ○ CMOS inverter Is-Vi   Select dimensions (width / length): [PMOS: ALFA PMOS 1, NMOS: ALFA NMOS 1 ▾]

Diode **characteristics:**

    ○ Diode I-V   Select diode: [Diode 1N4001 ▾]

BJT **characteristics:**

    ○ BJT npn i_C-V_{CE} Select BJT: [AC127 ▾]

    ○ BJT npn Gummel plot   Select BJT: [AC127 ▾]

2. Set up the voltage source(s):

Channel 1:			Channel 2:		
Start:	[0.0	V]	Start:	[2.5	V]
Stop:	[2.5	V]	Stop:	[3.0	V]
Number of steps:	[50]		Number of steps:	[3]	
Compliance:	[0.001	A]	Compliance:	[0.001	A]

3. Send to the server:

Do you want the results to open in a new window or in this window?

    ◉ New window   ○ This window

Push the "Submit" button to send the experiment to the server.

[Submit] [Reset]

Figure 1.21 Client window used with LabVIEW CGI script for selection of predefined experiments and setup parameters.

1.4.1 AIM-Lab CMOS Experiment

1.4.1.1 CMOS Assignment

- Go to http://nina.ecse.rpi.edu/shur/remote.
- Scroll down and push the button marked Connect to the measurement setup.

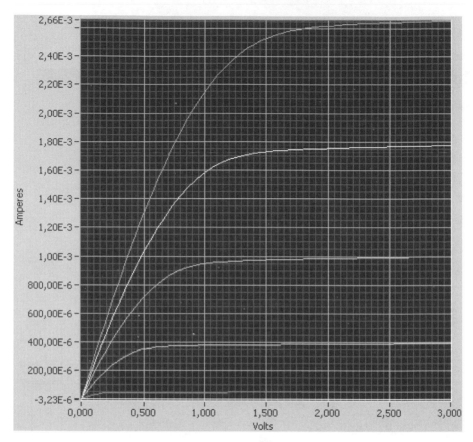

Figure 1.22 Client-side display of experimental NMOS characteristics using LabVIEW CGI solution.

- The Client window appears. Go to the Operation menu and choose Start experiment.
- The Channel Setup Dialog box appears. Select first NMOS Id-Vd characteristics.
- The Source Setup panel appears. This is the table where we enter the experimental parameters.

Note that the source and substrate of the transistor are connected to ground. Examples of client interactive pages in AIM-Lab are shown in Figure 1.2. The two transistors considered (NMOS and PMOS) have gate length $L = 2$ μm and width $W = 25$ μm.

1. Measure the NMOS I–V characteristics (i.e., drain current versus drain voltage). Choose different combinations of inputs to the parameter table.

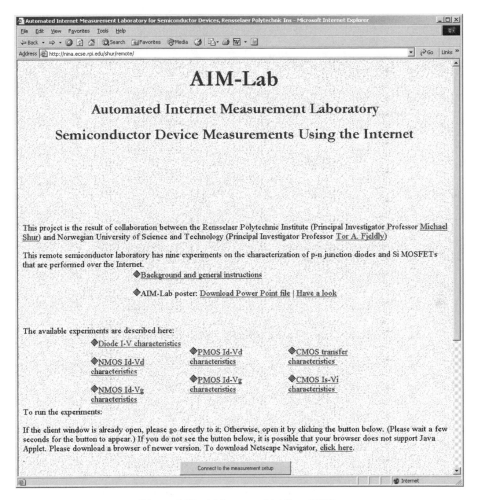

Figure 1.23   Main page of LAB-on-WEB.

2. Measure the NMOS transfer characteristics (i.e., drain current versus gate voltage) by selecting `NMOS Id-Vg characteristics` in the `Channel Setup Dialog` box. Choose a drain bias that corresponds to saturation for the range of V_{GS} used (see the characteristics in Figure 1.3). Since we are investigating a long-channel device, we may use the simple charge control model to approximate the measured dependence by the following equation:

$$I_{\text{sat}} = \beta_n V_{GT}^2 \quad \text{where} \quad \beta_n \equiv \frac{W\mu_n C_i}{2L} \quad \text{and} \quad V_{GT} \equiv V_{GS} - V_{Tn} \quad (1.1)$$

Here I_{sat} is the saturation current, μ_n is the electron mobility in the chan-

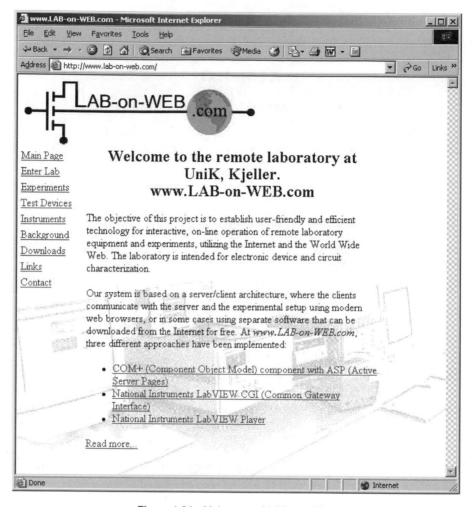

Figure 1.24   Main page of LAB-on-WEB.

nel, and C_i is the oxide capacitance per unit area. Plot $\sqrt{I_{\text{sat}}}$ versus V_{GS} and extract the value of β_n and the threshold voltage V_{Tn}.

3. Do a similar set of measurements for the PMOS. In this case the equations become

$$I_{\text{sat}} = \beta_p V_{GT}^2 \quad \text{where} \quad \beta_p \equiv \frac{W \mu_p C_i}{2L} \quad \text{and} \quad V_{GT} \equiv V_{GS} - V_{Tp} \quad (1.2)$$

Here μ_p is the hole mobility. Extract the value of β_p and the threshold voltage V_{Tp}. Note that in this case the currents and voltages have opposite sign of those of the NMOS.

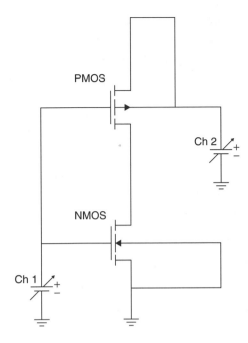

Figure 1.25   CMOS inverter. Input voltage V_{in} and voltage source V_{DD} are labeled Ch1 and Ch2, respectively.

4. Figure 1.25 shows a CMOS inverter circuit. Use Eqs. 1.1 and 1.2 to derive the following equation for the switching voltage of this inverter:

$$V_{sw} = \frac{V_{DD} + V_{Tp} + V_{Tn}\sqrt{\beta_n/\beta_p}}{1 + \sqrt{\beta_n/\beta_p}} \qquad (1.3)$$

Hint: Note that the currents through both transistors are the same. For $V_{in} = V_{sw}$ assume that the PMOS and NMOS are in the saturation regime.

5. Measure CMOS inverter transfer characteristics by selecting CMOS transfer characteristics in the Channel Setup Dialog box. Note that the CMOS inverter consists of the same two devices investigated above. Compare the switching voltages with those predicted by Eq. 1.3.

6. Use the circuit description below and insert the values for β_n, V_{Tn}, β_p, and V_{Tp} found above to simulate the CMOS inverter transfer characteristics. (Note that the SPICE parameter kp = μc_i for NMOS and PMOS are obtained from β_n and β_p and the device geometry L, W). SPICE uses the symbol vto for the threshold voltage. Compare with the experiments and comment.

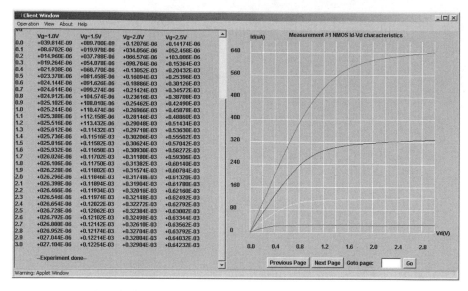

Figure 1.26   NMOS I_d–V_{DS} characteristics for $V_{GS} = 1$–2.5 V.

```
CMOS inverter
vdd 1 0 5
vin 2 0 0
m1 3 2 1 1 mp w=25u l=2u
m2 3 2 0 0 mn w=25u l=2u
.model mn nmos level=1 kp=xxxx vto=yyyyy
.model mp pmos level=1 kp=zzzz vto=uuuu
```

1.4.1.2 *Results*

1. First the NMOS I_d–V_{DS} characteristics are measured. An example is shown in Figure 1.26 for the default parameters in AIM-Lab.

2. For the parameter extraction, we also need the NMOS transfer characteristics (I_d versus V_{GS}). Such a characteristic is shown in Figure 1.27 for a drain bias of 4 V. The data in Figure 1.27 is used for calculating the dependence of $\sqrt{I_d}$ on V_{GS}, as shown in Figure 1.28. Fitting the above-threshold part with a straight line, we can extract the parameters β_n and V_{Tn} as follows:

$$|I_{\text{sat}}| = \beta_n(V_{GS} - V_{Tn})^2 \Leftrightarrow \sqrt{I_{\text{sat}}} = \sqrt{\beta_n}(V_{GS} - V_{Tn}) \tag{1.4}$$

which gives

$$\sqrt{\beta_n} \approx 1.3 \times 10^{-2}\ \Omega^{-1} \qquad V_{Tn} \approx \frac{7.02 \times 10^{-3}}{1.3 \times 10^{-2}} \approx 0.538\ \text{V} \tag{1.5}$$

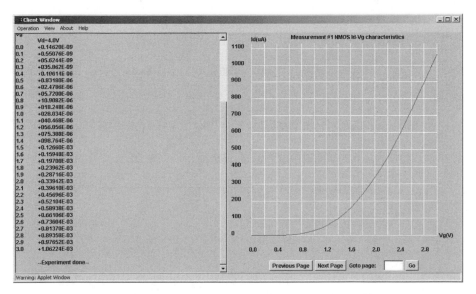

Figure 1.27 NMOS I_d–V_{DS} characteristics for $V_{GS} = 4$ V.

3. The PMOS device is characterized and the parameters are extracted similarly (see Figure 1.29). Here we find

$$\sqrt{\beta_p} \approx 9.553 \times 10^{-3} \ \Omega^{-1}$$

$$V_{Tp} \approx -\frac{4.348 \times 10^{-3}}{9.553 \times 10^{-3}} \approx -0.455 \text{ V} \tag{1.6}$$

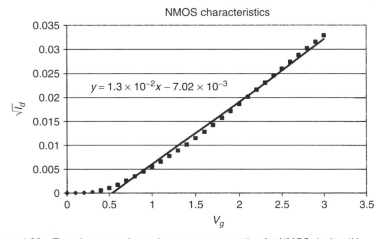

Figure 1.28 Experiment results and parameter extraction for NMOS device ($V_{DS} = 4$ V).

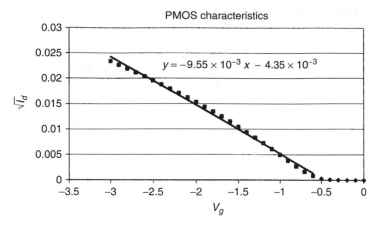

Figure 1.29   Experiment results and parameter extraction for the PMOS device ($V_{DS} = -4$ V).

4. Equating the square root of the NMOS and PMOS saturation currents at the switching point, we find

$$\sqrt{\beta_n}(V_{GSn} - V_{Tn}) = \sqrt{\beta_p}(-V_{GSp} + V_{Tp}) \tag{1.7}$$

Substituting $V_{GSn} = V_{in} = V_{sw}$, $V_{GSp} = V_{in} - V_{DD} = V_{sw} - V_{DD}$, and solving for V_{sw}, we obtain

$$V_{sw} = \frac{V_{DD} + V_{TP} + V_{TN}\sqrt{\beta_n/\beta_p}}{1 + \sqrt{\beta_n/\beta_p}} \tag{1.8}$$

From the parameters extracted above, we find

$$V_{sw} \approx \frac{V_{DD} + 0.2794}{2.366} \tag{1.9}$$

5. The measured CMOS inverter transfer characteristics are shown in Figure 1.30. Next, Figure 1.31 shows a comparison of the result from the above expression for V_{sw} with values obtained from Figure 1.30. Considering the accuracy of the fit and the precision of the V_{sw} readings obtained from Figure 1.30, Eq. 1.9 gives a good prediction of the CMOS inverter switching point.

6. The simulation results are compared with the experimental inverter transfer characteristics in Figure 1.32. Good agreement was obtained between experiments and simulations considering that only the simplest MOSFET SPICE model (level 1) was used. This shows that the PMOS and NMOS transistors with gate lengths of 2 μm used in this experiment can be considered as long-channel devices, for which the MOSFET level 1 SPICE model is applicable.

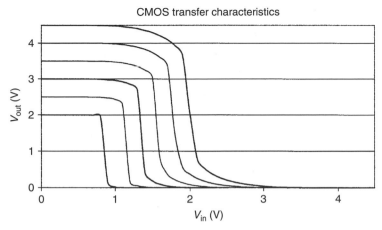

Figure 1.30 CMOS inverter transfer characteristics (V_{DD} ranges from 2 to 4.5 V with 0.5-V step).

1.4.2 LAB-on-WEB Subthreshold MOSFET Experiment

1.4.2.1 Background Very simplistically, the subthreshold regime of a MOSFET is considered an *off* state of the device, ideally blocking all drain current. In practice, however, there will always be some leakage current in this state owing to a finite amount of mobile charge in the channel and a finite injection rate of carriers from the source terminal into the channel.

The subthreshold current can be analyzed in a straightforward manner by considering a long-channel MOSFET with a gate length of a few micrometers or more. When such a device is biased in the subthreshold regime, the applied

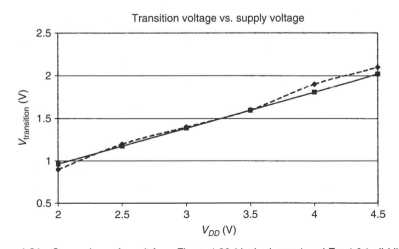

Figure 1.31 Comparison of result from Figure 1.30 (dashed curve) and Eq. 1.9 (solid line).

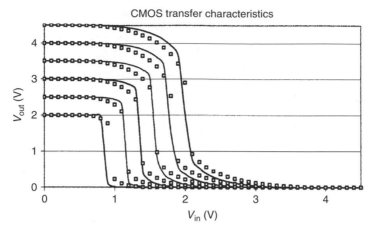

Figure 1.32 Comparison of experimental and simulated CMOS inverter transfer characteristics (experiments: solid lines; AIM-Spice simulations: symbols).

drain–source voltage will drop across the drain depletion zone of the channel. The remaining part of the channel is essentially at a constant potential (flat energy bands), where diffusion is the primary mode of charge transport. For this case, we easily find the following relationship of the MOSFET subthreshold drain current (see, e.g., Fjeldly et al., 1998):

$$I_{d,\text{sub}} \propto \exp\left(\frac{V_{gt}}{\eta V_{\text{th}}}\right)\left[1 - \exp\left(-\frac{V_{ds}}{V_{\text{th}}}\right)\right] \tag{1.10}$$

Here $V_{gt} = V_{gs} - V_T$ is the gate voltage swing, V_T is the threshold voltage [which separates the above-threshold behavior ($V_{gt} > 0$) from the subthreshold behavior ($V_{gt} < 0$)], V_{th} is the thermal voltage (0.026 V at room temperature), and η is an ideality factor related to the division of V_{gs} between the gate oxide and the semiconductor. We notice from Eq. 1.10 that $I_{d,\text{sub}}$ is an exponential function of V_{gt} and, when $V_{ds} > 2V_{\text{th}}$, it becomes independent of V_{ds} (saturation).

However, in modern-day MOSFETs, with submicrometer gate lengths, the subthreshold energy bands are no longer flat owing to the reduced geometry combined with a nonideal scaling of the devices. Moreover, the applied drain–source voltage will be distributed over the length of the channel, giving rise to a shift of the conduction band edge near the source end of the channel, as illustrated in Figure 1.33. This effect, known as drain-induced barrier lowering (DIBL), represents an effective lowering of the injection barrier for electrons from the source into the channel. Since the dominant injection mechanism is thermionic emission, the DIBL effect gives rise to a significant rise in the subthreshold current with increasing V_{ds}. This phenomenon can conveniently be

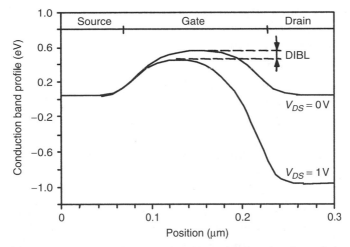

Figure 1.33 Conduction band profile at semiconductor–insulator interface of short *n*-channel MOSFET with and without drain bias. Figure indicates origin of DIBL (from Fjeldly et al., 1998).

described in terms of a shift in the threshold voltage. The dependence of V_T on the channel length L and the drain bias can be expressed as

$$V_T(L) = V_{T0}(L) - \sigma(L)V_{DS} \tag{1.11}$$

where $V_{T0}(L)$ describes the scaling of the threshold voltage at zero drain bias with L and $\sigma(L)$ is the channel-length-dependent DIBL parameter.

Clearly, the subthreshold current is very important since it has consequences for the bias and logic levels needed to achieve a satisfactory *off* state in digital operations. Hence, it affects the power dissipation in logic circuits. Likewise, the holding time in dynamic memory circuits is controlled by the magnitude of the subthreshold current. The DIBL effect tends to increase the power dissipation in the subthreshold regime and also increase the above-threshold current at high drain biases. However, well above threshold, the injection barrier is much reduced and the DIBL effect eventually disappears.

The objective of this assignment is to study the dependence of the threshold voltage on gate length and drain bias for the Alfa chip NMOS devices (see Section 1.2.1.1).

1.4.2.2 Subthreshold MOSFET Assignment

- Go to `http://www.lab-on-Web.com/`.
- Get acquainted with LAB-on-WEB by looking through the pages of the website.

- Look up NMOS Id-Vg (Drain Current—Gate Voltage) Character-istic under Experiments.
- Select Enter Laboratory.
- Choose SVG Graphics Interface.
- Under Auto Setup select the predefined experiment NMOS Id-Vg.
- Under Dimensions select NMOS 25/0.7.
- Enter measurement parameters in the Ch. 1 (Vg $\equiv V_{gs}$) and Ch. 2 (Vd $\equiv V_{ds}$) tables.
- Under Plot Setup select Logarithmic on the left vertical axis.
- Press Execute Measurement.

1. Measure the NMOS 25/0.7 I_d–V_g characteristics for V_{gs} between 0.3 and 1.3 V for the following values of V_{ds}: 0.05, 1.05, 2.05, and 3.05 V.
2. Extract the slope of the semilogarithmic (base 10) characteristics in the subthreshold regime (typically for $V_{gs} < 0.6$ V) and determine the ideality factor η in Eq. 1.10.
3. From the subthreshold region, determine the threshold voltage shift ΔV_T for the various applied drain biases. Use the characteristic for $V_{ds} = 0.05$ V as a reference. Plot ΔV_T versus V_{ds} and estimate the DIBL parameter σ.
4. Repeat the procedure for the NMOS 25/0.8, 25/1.3, and 25/1.6 and use the results to plot σ versus L. Discuss the results.

1.4.2.3 Results

1. Figure 1.34 shows the NMOS 25/0.7 I_d–V_g characteristics for the following values of V_{ds}: 0.05, 1.05, 2.05, and 3.05 V.
2. We find that the linear part of the plots in Figure 1.34 can be expressed, in agreement with Eq. 1.10, as

$$\log_{10}\left(\frac{I_{d,\text{sub}}}{I_0}\right) = \frac{V_{gs}}{\eta V_{\text{th}}} \log_{10}(e) \qquad (1.12)$$

where I_0 corresponds to the current at $V_{gs} = \eta V_{\text{th}}$. Selecting two arbitrary points in the linear part of the plot, we extract the following reasonable value of the ideality factor:

$$\eta \approx 1.5$$

3. The threshold voltage shift $\Delta V_T(V_{ds})$ is obtained from Figure 1.33 by measuring the horizontal shifts of the different characteristics using $V_{ds} = 0.05$ V as a reference, and ΔV_T versus ΔV_{ds} measured at

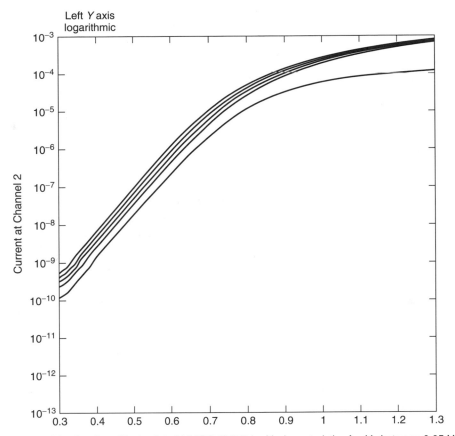

Figure 1.34 Semilogarithmic plot of NMOS 25/0.7 I_d–V_g characteristics for V_{ds} between 0.05 V (lower curve) and 4.05 V (upper curve), 1-V step. Vertical axis: drain current (A). Horizontal axis: drain–source bias (V).

$I_{d,\text{sub}} = 10^{-7}$ A is presented in Figure 1.35. The straight line corresponds to the following DIBL parameter:

$$\sigma = \frac{\Delta V_{Tp}}{\Delta V_{ds}} \approx 0.018$$

4. For increasing gate length, we will notice a decreasing shift of the subthreshold portion of the $\log(I_{d,\text{sub}})$ versus V_{gs} characteristics. Figure 1.36 shows the results measured for the gate length, $L = 1.6$ μm. Estimating the values of σ using only the curves corresponding to V_{gs} values of 0.05 and 4.05 V, we find the scaling of σ versus L plotted in Figure 1.37. The obtained scaling shows the expected trend. As L increases, the device approaches the ideal, long-channel, subthreshold behavior described by Eq. 1.10, where the DIBL effect becomes negligible.

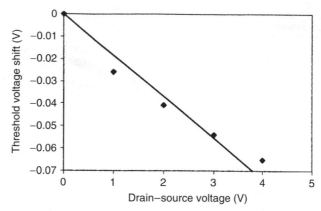

Figure 1.35   Plot of ΔV_T versus V_{ds} for NMOS 25/0.7. Experimental values (symbols) were obtained from Figure 1.34. Straight line corresponds to best linear fit. Slope of this line is DIBL parameter σ.

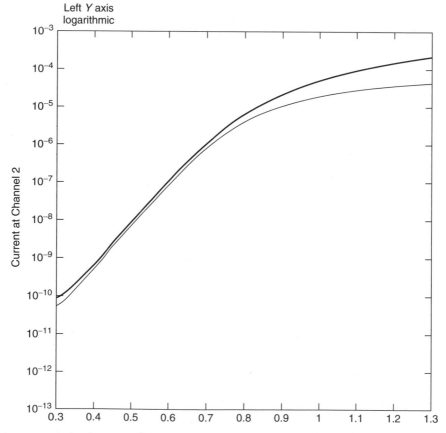

Figure 1.36   Semilogarithmic plot of NMOS 25/1.6 $I_d - V_g$ characteristics for V_{ds} between 0.05 V (lower curve) and 4.05 V (upper curve), 1-V step. Vertical axis: drain current (A). Horizontal axis: drain–source bias (V).

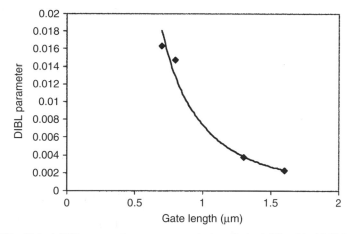

Figure 1.37 Plot of DIBL paramater σ versus gate length L of Alfa chip NMOS transistors. Experimental data are shown as symbols. Curve indicates the trend.

1.4.3   LAB-on-WEB Diode Experiment

1.4.3.1   Assignment

- Go to http://www.lab-on-web.com/.
- Get acquainted with LAB-on-WEB by looking through the pages of the website.
- Look up Diode I–V (Current-Voltage) Characteristic under Experiments.
- Select Enter Laboratory.
- Choose LabVIEW CGI Interface.
- Select the predefined experiment Diode I–V (diode 1N 4001).
- Enter measurement parameters in the Ch. 1 table under Set up the voltage source(s).
- Press Submit.

1. Measure the *I–V* characteristic between 0.05 and 0.7 V. Replot the curve in a semilogarithmic scale by copying and pasting the data in another application, such as Excel (note that you may have to replace periods by commas to be able to plot in some applications). Discuss the various regions of the curve.

2. Consider the semilogarithmic plot (base 10) for the voltage range between approximately 0.3 and 0.6 V. Extract the saturation current I_s and the ideality factor η from an extrapolation of $\log_{10}(I_s)$ to 0 V and from the slope, respectively. What does the value of η reveal about the current mechanism? What does the change of slope below 0.3 V signify?

3. Measure the current–voltage characteristic of the diode in the range −0.01 to +0.01 V. Use this curve to extract the ratio $I_s/(\eta V_{th})$ from the slope of the diode characteristic close to 0 V ($V_{th} = 0.026$ V is the thermal voltage assuming room temperature). Compare with the results from 2 and comment on the difference, if any.

4. Measure the reverse diode characteristic from 0 to −3 V. Comment on the voltage dependence of the reverse current.

1.4.3.2 Results

1. *Forward Bias.* Figure 1.38 shows the characteristic for diode 1N 4001 in a linear plot obtained using the LabVIEW CGI interface in LAB-on-WEB. The accompanying raw data were exported to Excel and are shown in a semi-logarithmic plot (base 10) in Figure 1.39. The straight line in this figure indicates the approximate fit to the "ideal" diode region for applied voltages V

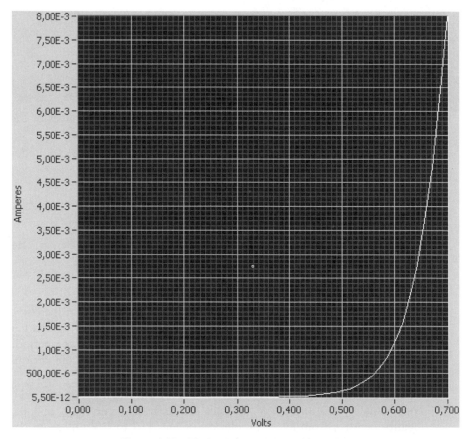

Figure 1.38   Diode characteristics in linear plot.

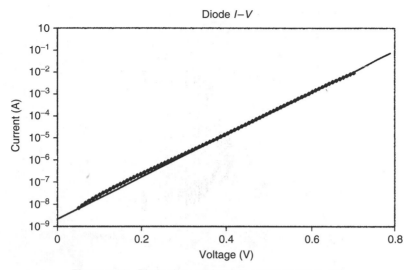

Figure 1.39   Diode characteristics in semilogarithmic plot.

between 0.3 and 0.6 V, given by the familiar diode equation

$$I = I_s \left[\exp\left(\frac{V}{\eta V_{th}}\right) - 1 \right] \qquad (1.13)$$

Here I_s is the saturation current, η is the ideality factor (usually between 1 and 2), and $V_{th} = 26$ mV is the thermal voltage at room temperature. Note that for the voltage range considered the exponential term will dominate in Eq. 1.13 since $V \gg \eta V_{th}$. At the lower voltage range, we notice a slight deviation from the ideal line resulting from excess current owing to recombination processes. Above 0.6 V, we see indications of a reduction below the ideal line owing to high-injection effects and the series resistance.

2. *Parameter Extraction.* The saturation current is found by extrapolating the straight line to the vertical axis at $V = 0$ V. The intercept gives $I_s = 2 \times 10^{-9}$ A. The slope of the curve can be obtained from

$$\log_{10}\left(\frac{I}{I_s}\right) = \frac{V}{\eta V_{th}} \log_{10}(e) \qquad (1.14)$$

Taking the derivative of $\log_{10}(I/I_s)$ with respect to V, we find the slope $\log_{10}(e)/\eta V_{th}$. From the curve, we obtain the slope 9.6. Using $\log_{10}(e) = 0.4343$, we find the ideality factor $\eta = 1.74$. This is a high number, indicating the presence of nonideal effects throughout the characteristic. A reduced slope below 0.3 V normally indicates the presence of recombination mechanisms.

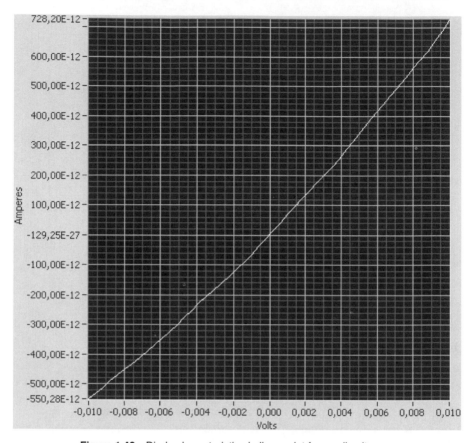

Figure 1.40   Diode characteristics in linear plot for small voltages.

3. *Small Forward Bias.* Figure 1.40 shows the diode characteristic measured for small voltages in the range from -0.01 to $+0.01$ V. For very small voltages, the exponential function of the diode equation can be expanded to first order to give $I = V I_s / \eta V_{th}$ and the slope becomes $I_s / \eta V_{th}$. From Figure 1.40, we find $I_s / \eta V_{th} = 6.5 \times 10^{-8}$ A/V. From 2, we find the corresponding value of 4.4×10^{-8} A/V. This indicates basically that the saturation current corresponding to the generation/recombination current is somewhat larger than that of higher voltages, as could be expected. However, the difference is not very significant, which is in agreement with the observation in Figure 1.39. The large ideality factor, even at the higher voltage range, may indicate that the diode is doped with recombination centers (e.g., to make the diode faster).

4. *Reverse Bias.* The diode characteristic for reverse bias is shown in Figure 1.41. We notice that this current does not really saturate with increasing reverse bias. Instead, it increases gradually. This behavior is related to the fact that the generation current is proportional to the width of the depletion zone. In fact,

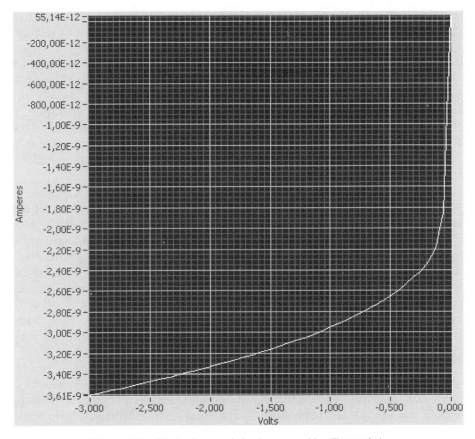

Figure 1.41 Diode characteristics in reverse bias (linear plot).

we should expect to see a square-root dependence of the generation current junction versus $(V_{bi} + |V|)$, with some adjustment for the deviation in the doping profiles through the diode from that of an abrupt junction.

1.5 EDUCATIONAL EXPERIENCE AND FUTURE PLANS

We have experience of the use of our remote laboratories as modules in the first-year graduate course Semiconductor Devices and Models I (SDM-1) at RPI and in the senior course Device Modeling and Circuit Simulation at UniK/NTNU. Both courses were offered as parts of a distance-learning program at the two institutions. The experiments included those discussed in the examples of Section 1.4 and additional experiments on bipolar transistors and photodiodes. Several of the experiments are available from both locations, with different devices, interfaces, and experimental details.

SDM-1 at RPI was a course for first-year graduate students and qualified seniors offered to both on-campus students and distance-learning students from General Motors, IBM, Pitney Bowes, and other companies. A total of 23 students were enrolled in the pilot course. All course materials were posted on the Web, and students did not have to remember any additional Universal Resource Locators (URLs). The experiments in AIM-Lab were used in the classroom to illustrate and reinforce the basic principles of the operation of the FET as well as to demonstrate some nonideal effects, which limit the FET performance. The ability to change the voltage and current ranges and to zoom in on certain features of the current–voltage characteristics was especially useful.

We also compared the measured data with FET models. This was done in two different ways. First, the students had to comment if the results of the measurements made sense given only the geometry of the device under test. They used order-of-magnitude values of the electron mobility and saturation velocity to roughly estimate the maximum device current and transconductance. This led to a fairly lively discussion in class. Only then were different FET models used, from a very simple constant-mobility model to the more sophisticated models, in order to try to fit the device characteristics. The students were also asked to provide a feedback and critique the user interface. Their comments and critique helped to improve the remote laboratory. It was a very positive and rewarding experience.

Similar experiences were gained at UniK/NTNU. Here, local and remote students sit at different campuses that are interconnected by a high-speed network, allowing us to stream two-way video over the Internet (IP). Overheads, SPICE simulations, and remote lab experimental demonstrations are also transmitted over the Internet using the NetMeeting application by Microsoft. This high-quality communication system allows projections on large screens to create a classroomlike environment also for the remote students, which seems to help in reducing the barrier for students to engage in two-way interaction.

The student lab assignments at UniK/NTNU are performed using LAB-on-WEB and occasionally AIM-Lab, as a mandatory module in the course SIE4090 Device Modeling and Circuit Simulation. This course is based on the text by Fjeldly et al. (1998), and the experiments are designed to illustrate and reinforce the subject matter being taught. The assignment typically consists of background theory, measurements, parameter extraction, circuit simulations, and discussion of results. The experiments concern basic properties and nonideal effects of devices and subcircuits, including $p–n$ diodes (see Section 1.4.3), bipolar junction transistors (diode characteristics, Gummel plots, current gain), and MOSFETs (see Sections 1.4.2 and 1.4.4). The objective is to provide future circuit designers with a deeper insight into the models used in computer-aided design (CAD), to appreciate the strengths and weaknesses of the models, to understand the limitations of the models, and to familiarize themselves with the meaning of the model parameters.

All in all, AIM-Lab and LAB-on-WEB did provide a new and very useful dimension to the courses, so much so that we want to expand the labs to include several additional experiments using more advanced CMOSs (see, e.g., Chapter 4 of this book), bipolar junction transistor (BJT) integrated circuits, and also alternative technologies such as Group III–V metal–semiconductor FETs (MESFETs) and heterostructure FETs (HFETs).

We also plan to start developing remote lab experiments on logic circuit elements based on the use of FPGAs (field programmable logic arrays) for use in introductory electronic and computer engineering classes. These classes often have large numbers of students (typically about 500 at NTNU), where the use of remote lab technology may ease severe logistic and cost problems encountered in conventional labs.

We envision an increased collaboration in the development, establishment, and maintenance of remote laboratories between universities. This way, resources will be better utilized, laboratories will be made accessible to less endowed schools (including schools in developing countries), students will be able to use more advanced instruments, and the number and quality of available experiments may be vastly expanded. RPI and UniK/NTNU have already enjoyed such collaboration for several years.

We also foresee the establishment of Internet access to major research and engineering laboratory facilities. In such a case, scientists or engineers may, for example, submit their sample by mail and have it mounted by local staff, whereupon they are enabled to run their experiments remotely without the need for time-consuming and costly travel. Another possibility is for device and systems manufacturers, vendors, or independent entrepreneurs to establish remotely operated test stations where new products can be tested and evaluated by potential customers from all over the world.

1.6 CONCLUSION

We have developed and investigated different system solutions for allowing remote clients (students) to perform real laboratory experiments via the Internet. We have established two physical laboratory sites, AIM-Lab at RPI in the United States and LAB-on-WEB at UniK/NTNU in Norway. Both sites provide experiments on semiconductor devices that are accessible to clients worldwide.

The AIM-Lab site at RPI is based on a TCP/IP communication solution that uses a Java applet on the client side. This is achieved by means of a Java virtual machine in the web browser that can download and execute Java code. LAB-on-WEB at UniK/NTNU presently offers several different system solutions based on modern web and instrument control technologies, such as COM+ and LabVIEW 6i, and the advanced functionalities of today's web browsers. COM+ has a well-structured and flexible development environment that has been widely adopted by the web community. It also performs useful tasks related to security, queuing, and logging. LabVIEW solutions are easily developed by means of the graphical program development utility. Presently, two solutions based on CGI scripts and the LabVIEW Player executable are implemented in LAB-on-WEB.

Here, we have discussed our remote lab systems in some detail and provided examples of lab assignments that are presently being used as integral parts of courses at our institutions. The experience so far has been both positive and encouraging, with valuable feedback received from other users worldwide. Eventually, based on the remote lab concept, courses and course modules within many disciplines of engineering and science may be offered to remote students located any place in the world, including students who otherwise would be precluded by distance and lack of resources. We also foresee future applications where major research and engineering facilities open for remote experimentation and where sophisticated on-line test stations are established to provide potential customers with the opportunity to perform remote testing of new products.

ACKNOWLEDGMENTS

This work was supported by grants from the Nordic Council of Ministers through Nordunet2, from the Rensselaer Strategic Initiative Program, and from strategic initiatives at the Norwegian University of Science and Technology and UniK—University Graduate Center.

REFERENCES

J. A. del Alamo, J. Hardison, G. Mishuris, L. Brooks, C. McLean, V. Chan, and L. Hui, "Educational Experiments with an Online Microelectronics Characterization Laboratory," *Proc. Int. Conf. on Engineering Education (ICEE 2002)*, Manchester, United Kingdom, pp. O102.1–7 (2002).

R. Berntzen, J. O. Strandman, T. A. Fjeldly, and M. S. Shur, "Advanced Solutions for Performing Real Experiments over the Internet," *Proc. Int. Conf. on Engineering Education (ICEE 2001)*, Oslo, Norway, Session 6B1, pp. 21–26 (2001).

S. Delmas Bendhia, F. Caignet, and E. Sicard, "A Test Vehicle for Characterization of Submicron Transistors and interconnects," *Proc. Second IEEE Int. Conf. on Devices, Circuits and Systems (ICCDCS 1998)*, Margarita Island, Venezuela, IEEE Catalog No. 98TH8350, pp. 69–74 (1998).

T. A. Fjeldly, M. S. Shur, H. Shen, and T. Ytterdal, "AIM-Lab: A System for Remote Characterization of Electronic Devices and Circuits over the Internet," *Proc. 3rd IEEE Int. Caracas Conf. on Devices, Circuits and Systems (ICCDCS-2000)*, Cancun, Mexico, IEEE Catalog No. 00TH8474C, pp. I43.1–6 (2000).

T. A. Fjeldly, J. O. Strandman, and R. Berntzen, "LAB-on-WEB—A Comprehensive Electronic Device Laboratory on a Chip Accessible via Internet," *Proc. Int. Conf. on Engineering Education (ICEE 2002)*, Manchester, United Kingdom, pp. O337.1–5 (2002).

T. A. Fjeldly, T. Ytterdal, and M. S. Shur, *Introduction to Device Modeling and Circuit Simulation*, John Wiley & Sons, New York, 1998.

G. Geoffroy, T. Zimmer, M. Billaud, Y. Danto, H. Effinger, W. Seifert, J. Martinez, and F. Gomez, "A Practical Course in a Virtual Lab," paper presented at the Twelfth EAEEIE Annual Conference on Innovations in Education for Electrical and Information Engineering, Nancy, France, (2001).

I. Gustavsson, "Laboratory Experiments in Distance Learning," *Proc. Int. Conf. on Engineering Education (ICEE 2001)*, Oslo, Norway, August, Session 8B1, pp. 14–18 (2001).

K. Lee, M. S. Shur, T. A. Fjeldly, and T. Ytterdal, *Semiconductor Device Modeling for VLSI*, Prentice-Hall, Englewood Cliffs, New Jersey, 1993.

H. Shen, Z. Xu, B. Dalager, V. Kristiansen, Ø. Strøm, M. S. Shur, T. A. Fjeldly, J. Lü, and T. Ytterdal, "Conducting Laboratory Experiments over the Internet," *IEEE Trans. Educ.*, Vol. 42, No. 3, pp. 180–185 (1999).

H. Shen, M. S. Shur, T. A. Fjeldly, and K. Smith, "Low-Cost Modules for Remote Engineering Education: Performing Laboratory Experiments over the Internet," *Proc. 29th ASEE/IEEE Frontiers in Education Conference (FIE'00)*, Kansas City, Missouri, TID-7 (2000).

M. Shur, *Introduction to Electronic Devices*, John Wiley & Sons, New York, 1995.

E. Sicard, "Introducing Microelectronics CMOS Design on PC," INSA editor, http://intrage.insa-tlse.fr/~etienne, 2000.

A. Söderlund and K. Jeppson, "The Remote Laboratory—a New Compliment in Engineering Education," paper presented at the International Conference on Engineering Education (ICEE 2002), Manchester, United Kingdom, Paper No. O102 (2002).

J. O. Strandman, R. Berntzen, T. A. Fjeldly, T. Ytterdal, and M. S. Shur, "LAB-on-WEB: Performing Device Characterization via Internet Using Modern Web Technology," *Proc. IEEE Int. Conf. on Devices, Circuits and Systems (ICCDCS 2002)*, Aruba, IEEE Catalog No. 02TH8611C, pp. I022.1–6 (2002).

C. Wulff, T. Ytterdal, T. A. Sæthre, A. Skjelvan, T. A. Fjeldly, and M. S. Shur, "Next Generation Lab—a Solution for Remote Characterization of Analog Integrated Circuits," *Proc. Int. IEEE Conf. on Devices, Circuits and Systems (ICDCS 2002)*, Aruba, IEEE Catalog No. 02TH8611C, pp. I024.1–6 (2002).

2

MIT MICROELECTRONICS WEBLAB

J. A. del Alamo, V. Chang, L. Brooks, C. McLean,
J. Hardison, G. Mishuris, and L. Hui
Massachusetts Institute of Technology, Cambridge, Massachusetts 02139

2.1 INTRODUCTION

In the teaching of microelectronic device physics, hands-on characterization of transistors and other devices substantially enhances the educational experience. Through a properly constructed laboratory experience, students can characterize real devices and compare their operation with the theoretical models presented at lecture. In this way, they can extract parameters and reflect on nonidealities of the devices and shortcomings of the models. Through experimentation, students also gain an appreciation for device characterization techniques and develop data manipulation skills. Additionally, an open-ended laboratory experience allows students to independently explore device behavior following their curiosity. Close contact with the real world is always a powerful motivator and helps students learn better. The same can be said of many other domains of engineering where hands-on laboratory experimentation plays a critical role in high-level education.

Yet, in spite of these advantages, conventional courses in microelectronic device physics often do not include a laboratory component. This is because of equipment, space, user training, safety, and staffing constraints that become nearly insurmountable for medium- and large-size classes. For example, a versatile experimental setup to obtain the current–voltage characteristics of micro-

Lab on the Web: Running Real Electronics Experiments via the Internet
Edited by Tor A. Fjeldly and Michael S. Shur
ISBN 0-471-41375-5   Copyright © 2003 John Wiley & Sons, Inc.

electronic devices consists of a semiconductor parameter analyzer (such as an Agilent 4155B) to which the device under test, such as a transistor or diode, is connected. A semiconductor parameter analyzer is a costly state-of-the-art tool that is widely used by device engineers in industry. In addition to its cost, the complexity of this tool represents a significant barrier to its deployment in a class. A typical manual spans hundreds of pages covering the many modes of operation that the instrument supports. Its front panel is covered with many keys, and many more softkeys appear in the display in its various modes of operation. In addition to this, the logistics of student training and scheduling of exercises, particularly working with limited space, laboratory staff, and equipment and large numbers of students, can become daunting. If less sophisticated equipment is used, more test stations can be established, but at the cost of more space and staffing costs and a diminished educational experience.

A remote laboratory that allows the characterization of microelectronic devices without the need for the user to be in front of the experimental setup solves many of these logistics concerns while largely preserving the educational experience. The use of web technology, in particular, allows the creation of such a laboratory while imposing minimum requirements on the remote user. In fact, many instruments, including semiconductor parameter analyzers, already allow for a great degree of computer control. Hence, it is conceivable to "transfer" that control to a remote user through, for example, Java applet technology. In this approach, clients equipped with a simple Java-enabled web browser can communicate with the instruments from anywhere at anytime.

There are several advantages to an on-line microelectronics device characterization lab:

1. The experimental setup can be made available at any time of the day and night. This allows students to conduct their measurements whenever they wish.
2. There are no special staffing requirements. Once the device is in place, no further staffing of the lab is required.
3. The system is nearly as flexible as the instrumentation itself. This means that no new programming is necessary whenever a different device or measurement routine is required.
4. There are no safety concerns. Students work from the safety of their homes or institutional computer clusters. No safety training is required to use the system.
5. Scarce instrumentation and lab space can be effectively shared by many students. The system queues requests and executes them in real time. Under most circumstances, students have the feeling of "owning" the entire measurement setup. Furthermore, spreading the cost of the equipment among many users allows the use of the very same state-of-the-art equipment that students will face in industry after graduation.
6. Training is manageable since students need only learn those instrument

functions that have been programmed in the software interface. This cuts down the size of the manual from many pages to just a few. The manual can be made available on-line.

These advantages apply to laboratories other than microelectronics device characterization labs. Several groups around the world have developed on-line laboratories for experiments in robotics, chemical processes, structural dynamics, and others (Taylor and Trevelyan, 1995; Henry, 1996; Ferguson, 1997; Shor and Bandari, 1998; Shen et al., 1999; Salzmann et al., 2000; Dalton and Taylor, 2000; Ertugrul, 2000; Ko et al., 2000, 2001a,b; Berntzen et al., 2001; see also University of Illinois at Chicago Interactive Systems Laboratory, http://iel.isl.uic.edu/; Internet Based Remote Control System, http://aitac.np.edu.sg/ibrcs/; WEAVE, http://weaveserve. egr.duke.edu:8080/weave/index.jsp; and Lab-on-Web at UniK— University Graduate Center in Norway, http://www.lab-on-web.com/).

Over the last few years, we have built a system at the Massachusetts Institute of Technology (MIT) that makes microelectronics device characterization over the Internet possible (del Alamo et al., 2002a,b). We call it the MIT Micro-electronics WebLab, or WebLab for short (http://weblab.mit.edu). The goal of the WebLab project is to deliver a rich laboratory experience to any conventional web browser. Today, web browsers are ubiquitous; therefore, through WebLab, students can take current–voltage measurements on transistors and other devices in real time from anywhere and at anytime.

This chapter describes the WebLab system and the lessons we have learned from its development and deployment. It is organized as follows. Section 2.2 describes the architecture of the WebLab system. The user interface and its operation are detailed in Section 2.3. Section 2.4 discusses the educational use of WebLab to date and the lessons learned. Recently, we have been working to add collaboration features to WebLab; Section 2.5 summarizes our experience to date. Section 2.6 summarizes our major findings and charts out the future of the WebLab project at MIT.

2.2 WEBLAB SYSTEM ARCHITECTURE

At its essence, the WebLab system consists of instruments to characterize mi-croelectronic devices and a group of computer hardware and software components that bring the laboratory experience onto the World Wide Web. Since its beginning in 1998, the WebLab project has emphasized continuous improvement and early educational deployment. This process has resulted in more than five versions of WebLab, running from version 1.0 (deployed in September 1998) to version 4.2 (deployed in September 2001). Each successive version benefited from MIT student feedback and, as a result, improved upon the previous version in accessibility, functionality, reliability, and quality of user interface. Across all these WebLab versions, the core architecture has not changed. This is a highly scalable architecture that makes it possible for WebLab to be deployed simultaneously to hundreds of students in different subjects. The scalability of WebLab derives from its minimal use of the server and the test equipment during client sessions. This is explained in this section.

WebLab delivers the subordinate sweep measurement functionality of an Agilent 4155B semiconductor parameter analyzer. In such a measurement, one or two channels of the Agilent 4155B stimulate the connected microelectronic device with a *sweeping* voltage or current, and the other ports either serve as constant voltage/current sources or voltage/current monitors. A port performs a voltage/current sweep by generating multiple voltage/current values in a specified range. The sweep begins at the *start* value and moves up by the *step* value at each time increment until the *stop* value is reached. At the end of a sweep, $\{[(\texttt{stop-start})/\texttt{step}]+1\}$ data points are collected. Through a sweep measurement, one can observe how the terminals of a microelectronic device respond to different levels of voltage/current stimulation.

To perform a measurement on the Agilent 4155B, the user defines a test vector that describes the measurement and executes the vector on the instrument. The test vector specifies the function of each channel of the Agilent 4155B and the values relevant to the function (for a more detailed description of the test vector, see Section 2.3.1 below).

In a WebLab session, the user prepares test vectors, executes them, views the obtained data, and downloads the data to the local machine. Under the WebLab architecture, test vector preparation is done through the WebLab Java applet on the client computer. The inputs are also validated on the applet, so invalid inputs are never submitted to the server, thereby reducing server traffic and increasing the availability of the lab equipment. Only when a valid measurement is submitted does the client establish connection with the instruments in the lab. The WebLab server queues the incoming requests until the instruments become available. The semiconductor parameter analyzer in WebLab executes a measurement in a few seconds (the exact length of time depends on the details of the measurement), so the queue is rarely long.

After measurement execution, the instruments return numerical results to the user through the server, and the server terminates the connection with the client, instantly making the lab available again to a different user. The display and processing of the results are carried out on the client machine. Throughout an entire session, the lab instruments are dedicated to any one user for a time that is typically on the order of tens of seconds, the actual time required by the instruments to execute a test vector. This design fully realizes the sharing potential of web-controlled labs that leads to high responsiveness, great access flexibility, and extremely low per-user costs.

WebLab started as a single-device system. This means that at any one time there is only one transistor or other microelectronic device physically hardwired to the semiconductor parameter analyzer. WebLab versions 1.0–3.1 have all been single-device architectures. In our educational experiments we found that the single-device architecture did not deliver enough flexibility and reliability.

The fact that different classes that work on different devices cannot use the system simultaneously limits the architecture's flexibility. Its lack of redundancy limits its reliability. When a device is blown out, which occurs frequently when large classes have assignments, the system remains down until the device is replaced. In order to correct these deficiencies, in June of 2000 we deployed a new multidevice version of WebLab (version 4.0) that allows up to eight devices to be hardwired to the system at any one time. A switching matrix is used to select the device under test. The device selection is specified by the user as part of the test vector. Hence, at this time, there exist two different (although very similar) architectures for WebLab: the single-device architecture (used in versions 1.0–3.1) and the multiple-devices architecture (used in versions 4.0 and up) that includes a few extra components. Both types of systems are being used today. This section describes both architectures.

2.2.1 Single-Device Architecture (WebLab Versions 1.0–3.1)

Figure 2.1 displays the key hardware and software components of the single-device WebLab system and the data path that links them together. The laboratory portion of the single-device WebLab system consists of a semiconductor device and a semiconductor parameter analyzer that carries out current–

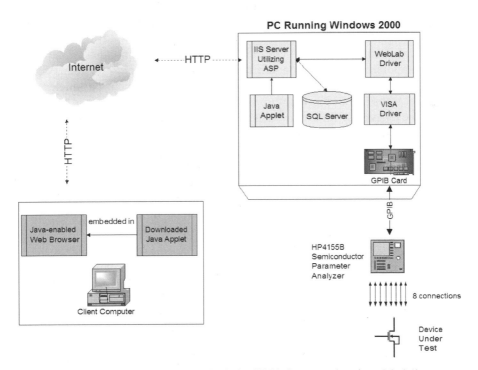

Figure 2.1 Architecture of single-device WebLab system (versions 1.0–3.1).

voltage measurements of the device. The semiconductor device—referred to as a device under test (DUT)—is the object of experimentation in student lab assignments. The system uses an Agilent 4155B semiconductor parameter analyzer—a state-of-the-art, industrial-strength instrument of broad use in the microelectronics world. It sends test currents and voltages through its connections to the DUT and measures the response of the connected device.

The Agilent 4155B semiconductor parameter analyzer has eight bidirectional channels: four SMUs (source/monitor units), two VMUs (voltage/monitor units), and two VSUs (voltage source units). The SMUs are the most versatile and commonly used channels. Each SMU can serve as a constant- or variable-current source or voltage source. When serving as a current source, the SMU monitors the voltage of the channel. When serving as a voltage source, it monitors the current running through the channel. VMUs are high-quality voltage monitors, and VSU channels are dedicated voltage sources.

The setup just described represents a standard microelectronics device characterization lab with substantial educational value in courses in microelectronics device physics, technology, and circuits. In WebLab, a series of software components cooperate with the hardware to make the laboratory a web-accessible lab. These components that reside on the WebLab server (a Windows 2000 server) form a data path between the laboratory equipment and any Java-enabled web browser on the Internet. In this path, data are converted back and forth between textual information, which can be read by the user, and GPIB commands, which are control commands that are understood by the instruments.

A key component of WebLab, the Java applet, is downloaded to the user's client computer. The applet contains the user interface of the system. Via the controls on the user interface, the user can enter measurement parameters that the applet transmits to the WebLab server. Before each transmission, the applet checks the test vector for common errors that would render it not executable by the Agilent 4155B and refuses to submit it until all such errors have been corrected by the user. When the test vector passes all the rules in the applet's error-checking procedure, the applet submits it to the WebLab server using an HTTP connection. On the WebLab server, one of a set of ASPs receives the uploaded vector through Microsoft Internet Information Services (IIS). IIS includes a request queue which the WebLab architecture fully utilizes. If another user is executing a measurement when the request is received, the request is queued on IIS until the instruments become available.

When the request gets off the queue, the ASP wraps the data into method calls to the third software component, the WebLab driver. This Visual-Basic-based driver places the user's request into method calls to the fourth software component, a commercially produced driver called VISA. The VISA driver then translates the Visual Basic commands into GPIB commands. GPIB is the instrument control language used by the Agilent 4155B. Through a commercially available GPIB interface card, the WebLab server transmits the GPIB method calls to the parameter analyzer. The user's measurement request is thus relayed to the lab equipment through this chain. In the background of this information flow, a fifth component, a database, records and supplies user information and transaction data by communicating with the ASP.

Upon receiving the request, the parameter analyzer performs the specified measurements and responds to the GPIB interface card with the results. Through the same path on which the request traveled to the lab equipment, the measured data travel back to the client applet, which then graphs them for the user. Since HTTP connections only persist through one request–response cycle, the connection is closed by the server as soon as it returns the measurement data.

2.2.2   Multiple-Device Architecture (WebLab Versions 4.0–4.2)

WebLab versions 4.0 and higher use a system architecture that includes multiple DUTs. This architecture is very similar to the single-device architecture described previously but with two significant differences. First, the multiple-device architecture includes an Agilent E5250A switching matrix. Second, because of the addition of the switching matrix, eight DUTs, instead of one, are accessible for testing at any one time.

Figure 2.2 displays the multiple-device architecture. In this system, the switching matrix routes signals between the Agilent 4155B parameter analyzer and the eight DUTs. Like the parameter analyzer, the switching matrix can be controlled remotely through its GPIB connection to the WebLab server. In this system, the user can remotely select one of the eight connected DUTs as the device to be measured. By means of the same data path described in Section 2.2.1, the user's selection request reaches the switching matrix in the form of GPIB commands. The switching matrix then connects the parameter analyzer with the selected device. When the parameter analyzer performs measurements, the switching matrix ensures that all input–output signals run between the parameter analyzer and the selected device. This multiple-device WebLab system offers much more flexibility and reliability than the single-device system. However, the switching matrix is an expensive purchase and may not be affordable to all institutions. Therefore, both single- and multiple-device systems continue to be developed at MIT.

In the latest version of the multiple-device system—version 4.2—a thermometer was added to the system architecture. This enables the WebLab user to obtain the ambient temperature surrounding the DUTs. This is a useful addition with significant educational impact because the behavior of microelectronics devices is often influenced in a large way by their surrounding temperature.

2.3   WEBLAB USER INTERFACE

The WebLab user interface has two main components: the *system management pages* and the *Java applet*. The system management pages consist of a group of ASPs that perform tasks, such as user authentication and resource man-

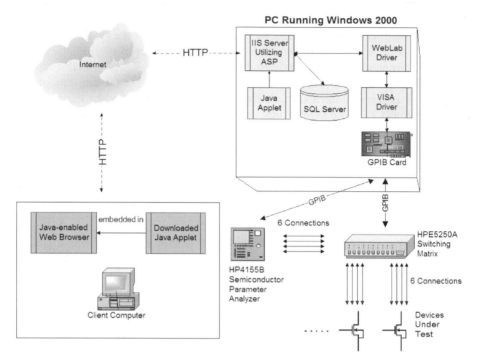

Figure 2.2 Architecture of multiple-device WebLab system (versions 4.0–4.2). Note that there are only six connections between each DUT and the switching matrix because the switching matrix only offers 48 ports.

agement, that are not directly related to the control of the lab instruments. When the user first connects to WebLab at http://weblab.mit.edu, an ASP greets the user and prompts for a username and a password (see Figure 2.3). This ASP submits the username and password to the database on the WebLab server for verification. If the user is authenticated as an approved WebLab user, he or she is directed to another ASP that provides access to the manual plus the other key component of the user interface—the Java applet. If the user is authenticated as an administrator of WebLab, he or she is directed to an ASP that also provides access to a whole set of ASPs that enable the remote management of the WebLab system (see Figure 2.4).

These remote management features were only recently added to WebLab in version 4.2. The initial development efforts toward WebLab had focused on delivering remote control of the lab to remote users. In our more than three years of experience in administering WebLab, we have realized that delivering the management of the lab to the system administrator over the Web is crucial for the effective upkeep of the system, especially as the number of WebLab users and the number of courses and experiments in which it is used increases. Through the features in the remote management ASPs, user infor-

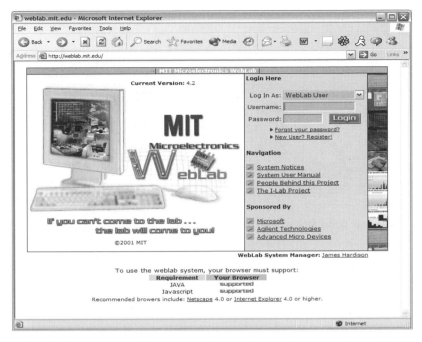

Figure 2.3 WebLab version 4.2 home page (`http://weblab.mit.edu`), an ASP that authenticates the user.

mation, login history, and session history—almost any data in the WebLab database—can be accessed and modified. Since their implementation, the new features have greatly enhanced our efficiency in administering the system.

The Java applet is the core component of the WebLab user interface. The applet provides a user interface that was inspired by the control panel of the Agilent 4155B parameter analyzer; through the controls on the applet, the parameter analyzer can be remotely operated. The user interface consists of two frames: `Measurement Specification`, which receives user input, and `Measurement Results`, which graphs the results of the experiments the user defines. The features of these frames are explained in the following sections.

2.3.1 `Measurement Specification` Frame of Java Applet

The `Measurement Specification` frame has three panels (see Figure 2.5). At the top of the frame lies the `Channel Definition` panel in which eight rows of input fields are displayed, representing the eight channels of the parameter analyzer (four SMU channels, two VSU channels, and two VMU channels). The input fields of the `Channel Definition` panel are as follows (from left to right):

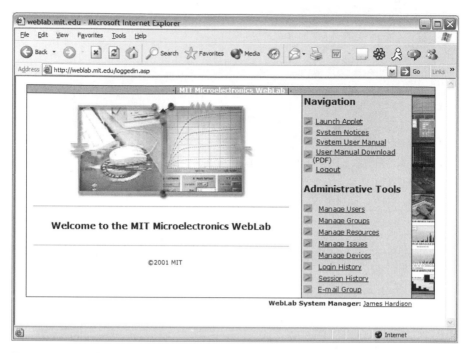

Figure 2.4   Main WebLab portal. This ASP is returned to a user after being authenticated as a WebLab administrator. Links in top half of page provide access to applet and user manual. They are available to all users. Links in bottom half of page provide administration features. They are only available to users with administrator privileges.

- VName and IName: Each unit has a voltage and current associated with it, and their names are chosen by the user.
- Mode: The mode of a unit can be V, I, or COMM. For a SMU, the mode determines what type of source it becomes. In V mode, the SMU becomes a voltage source and a current monitor. In I mode, it becomes a current source and a voltage monitor. In COMM mode, the unit is set to a common ground. Unlike SMUs, VSUs and VMUs can only function in one role—VMUs as voltage monitors and VSUs as voltage sources—so these types of units are always in V mode.
- Function: The function of a unit determines whether and when the values taken by a unit in the I or V mode get swept. The variable selected as VAR1 is the primary sweep variable that sweeps through its specified range multiple times. The variable selected as VAR2 is the secondary sweep variable that only sweeps through its specified range once. Every time the VAR1 variable completes a sweep, the VAR2 variable increments its value by one step size. This cycle repeats until the VAR2 variable has completed one sweep. If the function is set to CONS, the unit remains at a constant value that can be specified.

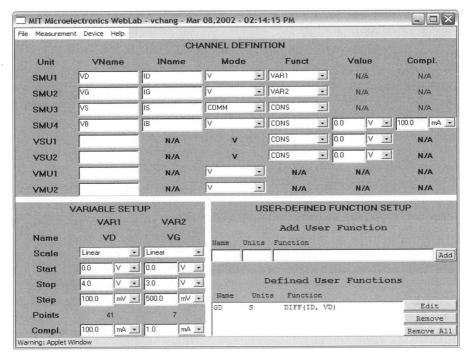

Figure 2.5 Measurement Specification frame of WebLab user interface. Test vector has been programmed to obtain output characteristics of *n*-channel MOSFET shown in Figure 2.6. Results of measurements shown in Figure 2.7.

- Value: This field only applies to units in CONS mode. The value in this field sets the value at which the unit is held constant.

- Compliance: Similar to the Value field, this field only applies in CONS mode. The compliance sets a maximum limit for the current or voltage that can be collected by this unit; any signal that exceeds this limit is not allowed to pass through the unit. This is done to protect the connected DUT.

When a unit has been selected as VAR1 or VAR2, a set of input fields for the unit appears in the Variable Setup panel located on the lower left corner of the Measurement Specification frame. In this panel, the user can specify the sweeping parameters of the selected units:

- Scale: This field determines the scale in which the variable is to be swept. For the variable selected as VAR1, there are four choices. In the linear scale, one data point is taken after every step and the size of the step is set by the user. In the log 10, log 25, and log 50 scales, 10, 25, and 50 points are taken every decade, respectively; the points are equally spread out logarithmically. The variable selected as VAR2 can only be swept in the linear scale.

- `Start`: This field determines the starting value of the sweep.
- `Stop`: This field determines the stopping value of the sweep.
- `Step`: If the linear scale is chosen for this unit, the `Step` field determines the step size between each data point. In any of the logarithmic scales, this field does not apply.
- `Points`: The value of this field represents the number of data points that will be taken during the sweeping measurement. It is automatically calculated by the applet based on the previous inputs.
- `Compliance`: Similar to the `Compliance` field in the `Channel Definition` panel, this value limits the amount of current (in V mode) or voltage (in I mode) this unit will allow. This limit is set to protect fragile devices.

It is often helpful to compute in real time functions of the data that are being collected. The Agilent 4155B makes it easy to do so through the user-defined functions. The WebLab user interface captures this feature in the `User-Defined Function Setup` panel located in the lower right corner of the `Measurement Specification` frame. For example, if a user wanted to analyze the square root of a variable and its relationships to other variables, he or she can define a user function that calculates the square root of the variable and see it graphed against other collected variables.

A user function consists of one or more data variables connected by operators defined by the Agilent 4155B. Defined operators include simple ones such as $+$, $-$, $*$, and $/$ and more complex ones such as AVG (average), SQRT (square root), and DIFF (derivative). User-defined functions can be nested inside one another. These functions are computed in real time by the Agilent 4155B. In WebLab, the results are returned to the Java applet along with measured data.

A menu bar that lies at the top of the `Measurement Specification` frame provides crucial controls of the WebLab system. Three menus (four menus in the multiple-device versions) exist in the menu bar:

- The `File` menu gives the user options to *save* or *load* a setup. All the values currently in the `Measurement Specification` frame can be captured and saved in the database on the WebLab server under a user-chosen setup name. After the setup is saved in the database, it can be loaded at any time; when loaded, all the captured values would be inserted into the proper fields on the frame. The load setup and save setup features are of great convenience to the user.

- The `Measurement` menu provides two options. `Show Terminal Configuration` brings up a diagram of the current device and its con-

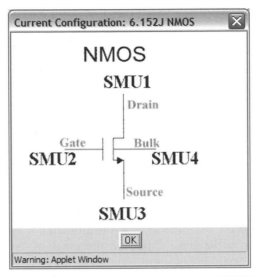

Figure 2.6 Terminal configuration of *n*-channel MOSFET.

nections with the ports of the Agilent 4155B (see Figure 2.6). Run Measurement is the most important option on this menu. When selected, the applet checks the inputs for errors. If the inputs are free of known errors, the applet launches a frame that prompts the user to select which of the defined variables—including all of the defined channels and user-defined functions—the user wants to have measured. After the selections have been made, all the information on the Measurement Specification frame is submitted to the WebLab server and the requested measurement is carried out.

- The Devices menu appears only in the multiple-device versions of WebLab. It lists all of the DUTs connected to the switching matrix. It is from this menu that the user selects the desired device. At any moment, one (and only one) device must be selected.

- The Help menu gives the user access to the user manual.

The Channel Definition, Variable Setup, and User-Defined Function Setup panels of the Measurement Specification frame collect measurement specification data from the user, and the menu options give the user the ability to manipulate the data and the frame. After the user submits data for measurement by selecting the Run Measurement menu option, the lab instruments execute the measurement and return the results to the Java applet. The results are displayed in the Measurement Results frame, which is described in the next section.

2.3.2 Measurement Results Frame of Java Applet

The Measurement Results frame contains a two-dimensional graph canvas that graphs data points sent to the client by the WebLab server. Three axes can be specified for the graph canvas: the X axis, the Y_1 axis (the primary Y axis, located on the left), and the Y_2 axis (the secondary Y axis, located on the right). Each axis can represent one of the variables that have been measured. The X axis and the Y_1 axis always have variables selected for representation. If a variable is selected for the Y_2 axis, the graph of Y_2 vs. X can be simultaneously displayed on the graph canvas. Each axis is assigned a control panel on the frame. Through this panel, the user selects the variable to be represented and the scaling parameters of the axis (maximum value, minimum value, unit, and scale).

When data points from a measurement are received by the applet, the Measurement Results frame is automatically launched. After selection of the variables to be graphed, the Auto Scale button on the frame allows the applet to automatically scale the graph based on the range of the data. Autoscaling usually produces a well-formed graph. The Accent Points checkbox lets the user view the actual data points in the graph by marking them with small, open circles. Figure 2.7 exhibits a graph that was produced after autoscaling and accenting the data points. If autoscaling does not generate exactly the graph that is desired, the user can also specify the axes' parameters manually through their respective control panels.

Besides displaying measurement results graphically, the Measurement Results frame also allows the user to view the data in numeric form. Through the See Data button on the frame, the applet launches another frame that displays every numeric data point received from the WebLab server for each measured variable (see Figure 2.8). The user can also download the numeric data to a file on the client machine by clicking the Download Data button. This gives the user the ability to manipulate and analyze the data in other programs, such as Excel or MatLab. The high degree of flexibility of the Measurement Results frame contributes great educational value to the WebLab system.

2.3.3 Example Interaction

This section presents an example WebLab interaction that illustrates how all of the features of WebLab described in the previous section work together. Bob, an undergraduate student in a microelectronics device physics course, is the main character of this example. His actions demonstrate a typical interaction with the system.

To begin a WebLab session, Bob starts a web browser on his computer and directs it to http://weblab.mit.edu. An ASP prompts him for his username and password (see Figure 2.3). He enters them and submits them to the server. The database verifies Bob as a WebLab user and directs him to an ASP through which he launches the Java applet.

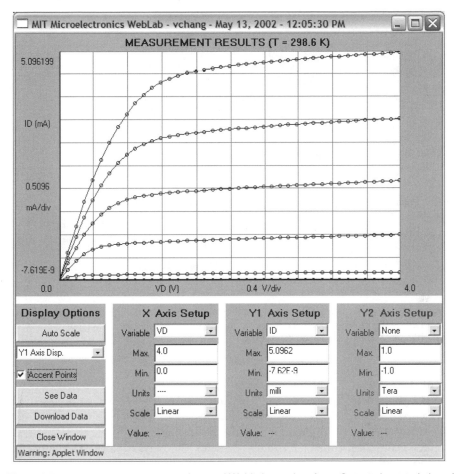

Figure 2.7 Measurement Results frame of WebLab user interface. Output characteristics of *n*-channel MOSFET is shown in frame. This graph was automatically obtained after using Auto Scale feature of frame.

Immediately, the Measurement Specification frame appears on Bob's computer screen. Bob's assignment is to obtain the output characteristics of a NMOS transistor. First, he clicks on the Devices menu and selects the NMOS transistor from the list of eight DUTs. He then clicks on the Measurement menu and selects the Show Terminal Configuration option to view the configuration diagram of the DUT (see Figure 2.6). Using the diagram as reference, Bob specifies the four SMU channels in the Channel Definition panel, as shown in Figure 2.5. VD (the voltage in SMU1, the drain of the device) is selected as VAR1 and VG (the voltage in SMU2, the gate of the MOSFET) is selected as VAR2. SMU3, the source, is set to ground, and SMU4, the body, is specified as a constant-voltage source (for this measure-

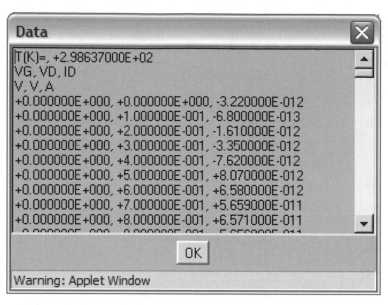

Figure 2.8 Numeric results of measurement displayed in a frame after user clicks See Data button on Measurement Results frame. Data in this figure correspond to output characteristic of *n*-channel MOSFET displayed in Figure 2.9.

ment it is set at 0 V, but this value can be changed). After selecting VAR1 and VAR2, Bob defines the sweeping parameters for those two variables in the Variable Setup panel. Bob is also interested in measuring the output conductance of the device; therefore, in the User-Defined Function panel, he defines a function called GD that represents the derivative of ID against VD (see Figure 2.5).

Once everything is defined to Bob's satisfaction, Bob clicks on the Measurement menu and selects the Run Measurement option. The applet checks Bob's inputs and discovers an error, so it rejects Bob's submission and launches an error message. Bob corrects the error and tries to submit the setup again, but he fails once more because the applet finds another error. This error correction cycle repeats until Bob has fixed all of the common errors for which the applet checks. Finally, no error message appears, and Bob is allowed to submit the measurement for execution. A frame that lists all of the defined variables appears on the computer screen and prompts Bob to select the variables he wants to have measured. Bob selects VD, ID, and GD and submits the measurement.

Within a few seconds, the Measurement Results frame appears with VD and ID already selected as the X and Y_1 variables, and the graph of Y_1 versus X is displayed on the graph canvas. Assuming that the axes are scaled improperly, Bob clicks the Auto Scale button, and instantly, the axes are rescaled to show the result. By checking the Accent Points checkbox, the data points in the graph become accented with circles. Figure 2.7 shows the resulting graph. Bob then selects GD as the variable for the Y_2 axis and autoscales again. Now, the graph canvas simultaneously displays the output conductance and the previous graph.

Bob is pleased with the graphs he has viewed, but he wishes to analyze the measurement results in Excel. He clicks on the See Data button to see the numeric data (see Figure 2.8); then he clicks on the Download Data button to save the data into a local file. He can now import the downloaded data file into Excel and perform numerical analysis on the data.

Bob has obtained all the results he wants. To avoid having to specify the same measurement again, he selects the Save Setup As option in the File menu. A frame is launched that prompts him to enter the name under which he wishes to save the setup. Bob enters the name NMOSChipTest and saves the setup; from now on, the setup can be retrieved from the database under that name. Having obtained the results he wanted and saved the setup for future use, Bob exits WebLab.

During Bob's interaction with WebLab, almost every one of the features on the user interface is used. This illustrates the practicality in the design of the Java applet. The applet encompasses many features, but all the features are frequently useful to the user.

2.3.4 Analysis of Accessibility and Scalability

One of the main benefits of an on-line laboratory is its ability to be efficiently shared among multiple users. Therefore, high accessibility and scalability are very important goals for WebLab's design. This section analyzes the system in these two dimensions.

Because WebLab places very few demands on the user's computer, it offers wide accessibility. It requires from the user no other software but a conventional web browser, which almost all PCs have. The WebLab applet is a very lightweight download of approximately 80 kB, so it places small requirements on the client's Internet connection. A computer connected to the Internet through a slow dialup connection with a transfer rate of 14.4 kbps can download the applet in about 45 sec. Today, most university campuses provide broadband Internet connections with transfer rates higher than 1.5 Mbps. Computers thus connected can download the applet in less than 1 sec. All communications between the applet and the server during a WebLab session are very lightweight, consisting of only textual data transfer of a few hundred bytes. We have run measurements from computers as far from MIT as Singapore, and the network latencies in applet download and measurement execution have always been found to be negligible.

WebLab was designed with scalability as a priority. As described earlier in Section 2.2, our design minimizes the amount of time the lab instruments are occupied for unproductive tasks. The only time in a measurement during which a user occupies the lab instruments is during the execution of the test vector, which only takes a few seconds. The more time-consuming tasks of test vector preparation and measurement result manipulation are performed off-line. Figure 2.9 illustrates this idea of minimizing instrument use.

During the fall term of year 2000, we analyzed WebLab's scalability during a week in which 75 students in an undergraduate subject on microelectronics devices and circuits at MIT had to carry out a WebLab assignment (for more information regarding our educational experiments, see Section 2.4 below). For this assignment, a total of 1237 measurements were submitted. Since the applet catches most of the errors on the client side, 92% of the requests received by the Agilent 4155B were valid test vectors and were executed. In other words, only 8% of the requests occupied instrument time with unproductive tasks.

In those experiments, we also found that the average execution time of a measurement (i.e., the amount of time the execution occupies the WebLab instruments) was 16 sec. Using that value, we estimate that, in steady state, WebLab can handle up to 225 requests of this kind per hour, or 37,800 requests per week. Figure 2.10 shows the hourly distribution of requests the system received during the studied week. In the peak hour of usage, the system received 99 requests, only 44% of its steady-state capacity. Of these 99 requests, over 70% of all requests were executed immediately after being received by the server, about 25% had to wait for one job to be finished, and 3% had two jobs ahead of them—the queue had hardly been used. Even at its busiest hour, WebLab still had plenty of capacity available. The distribution in Figure 2.10 illustrates typical college student behavior—system usage was heavily concentrated on the night before the assignment due date. While the peak hour on that night received 99 requests, the heaviest hour of usage observed on the other days received only 25 requests. This indicates that if different classes stagger their usage times so that their peak usage days do not overlap, the current implementation of WebLab can handle many more students from multiple classes.

Our discussion has shown that WebLab can be easily accessed from any location on the Internet, and it can simultaneously accommodate multiple classes involving hundreds of students. Compared to a traditional microelectronics device characterization lab, which can only serve a few small groups of students who can physically work in the lab, WebLab has far higher accessibility and scalability.

2.4   EDUCATIONAL USE OF WEBLAB

WebLab was first deployed for education at MIT in the fall of 1998. Since then the majority of WebLab usage has been by MIT students enrolled in MIT courses, but we have also carried out educational experiments in cooperation with National Singapore University and Compaq Corporation. Figure 2.11 summarizes the locations of a series of experiments carried out in the fall of 2000. These and other experiments will be described in this section.

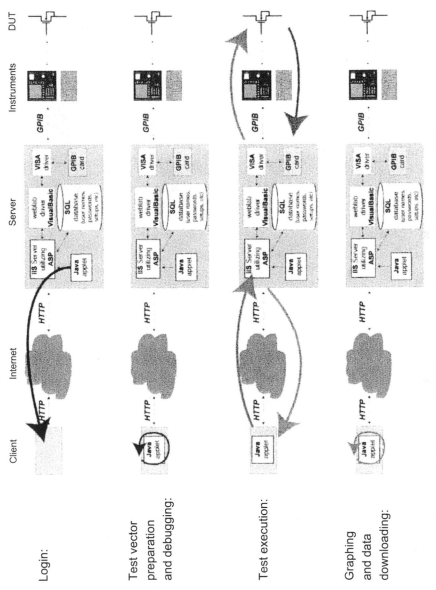

Figure 2.9 Sequence of subtasks in WebLab measurement. Only subtask that occupies lab instruments is test execution. This means that each measurement only occupies the instruments for a few seconds.

67

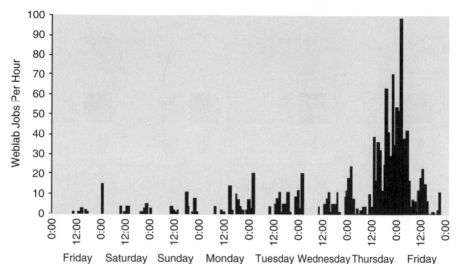

Figure 2.10   Hourly distribution of WebLab jobs received from October 13, 2000 to October 20, 2000 in undergraduate MIT subject on microelectronics devices and circuits with 75 students. Assignment went out at 2 PM on October 13, 2000 and was due at 2 PM on October 20, 2000.

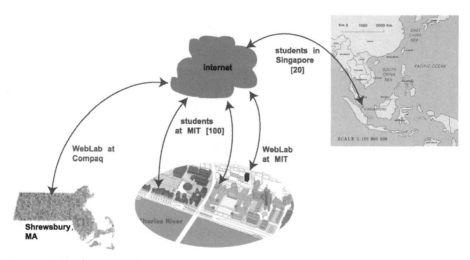

Figure 2.11   Summary of our educational experiments with WebLab in fall of 2000. During this semester, we deployed WebLab in various ways to MIT, National University of Singapore, and Compaq Corporation in Shrewsbury, Massachusetts. In the Compaq experiment, a copy of WebLab was installed on site at Compaq.

Over the last four years, WebLab has been used in three different courses: a junior-level subject on microelectronics devices and circuits at MIT ("6.012"), a graduate subject on microelectronics device physics at MIT ("6.720"), and a graduate-level subject on electronic materials taken by students in Singapore as part of the Singapore–MIT Alliance ("SMA5104"). None of these subjects previously included a lab component. This section summarizes the educational experiments that we have performed and what we have learned from them. It also briefly describes typical laboratory assignments that have been performed.

2.4.1 Educational Value of WebLab

In our educational experiments, we have found that there are three aspects to the educational experience associated with WebLab exercises. These are the construction of the test vector, management of the data display, and off-line data manipulation.

There is a great deal of educational value in the preparation of the WebLab test vector. Of the many test vectors that one could prepare, only a few respond to the specifications of the exercise, so the students are forced to pay detailed attention to the precise requirements of the exercise. Configuring the test vector also brings to the fore issues of measurement range, data point distribution, measurement speed, and device compliance. Additionally, the students are made aware of how the instrument actually carries out the measurements. The `Measurement Specification` frame of WebLab has been constructed to preserve a substantial portion of the experience associated with hands-on operation of the instrumentation. Since WebLab gives the students access to the internal functionality of the instrument, they are not spared from opportunities to prepare erroneous test vectors that do not respond to the assignment or that are not executable. Substantial learning takes place in the debugging of the test vector. The flexible nature of the `Measurement Specification` frame allows students to follow their curiosity and explore other modes of operation of the DUTs that go beyond the specified assignment. This is known to be one of the most educationally meaningful aspects of the laboratory experience in engineering education that WebLab captures well.

The second component of the WebLab educational experience concerns the data display. The `Graphics` frame of WebLab allows the user to easily select which variables to graph in three different axes (one abscissa and two ordinate axes), whether the scales are linear or logarithmic, and the range of all the scales. This graphing flexibility forces students to think about the optimal way to display the data and to study in detail the standard graphing formats followed in the microelectronics world.

The third educational aspect of the WebLab experience is off-line data manipulation. A download button on the Java user interface allows the exporting of the obtained data to a file in a format that is easily portable to many standard data analysis software tools. The student uses his or her favorite software package to further process the data, extract parameters, build simple models, and compare their predictions with the acquired data. One of our most significant findings is the clumsiness of most students in data manipulation. We discuss this below.

2.4.2 Educational Experiments

Since its first deployment, WebLab has been used at MIT in two main capacities. First and foremost, WebLab has been incorporated in homework and device characterization projects that students are asked to complete over a period of one to two weeks. Second, WebLab has been used in lecture to illustrate device operation while the relevant theoretical material is being presented. In the past, before WebLab, we performed these demonstrations by rolling out the actual Agilent 4155B to lecture with a camera directly over the front-panel display of the instrument. This camera captured the image and sent it to a wide screen in the auditorium. WebLab greatly simplifies the time-consuming procedure of giving live demonstrations since many classrooms these days come equipped with Internet connections and computer projection systems.

It is the use of WebLab in homework and laboratory projects where its greatest educational value lies. WebLab has enjoyed enormous success in this capacity. Since its first deployment in the fall of 1998, WebLab has been used in remote laboratory assignments by over 600 students. Figure 2.12 graphs the use of WebLab per calendar year. Just in the fall 2001 semester, over 150 students used WebLab in three different subjects simultaneously. The next subsections describe in more detail the use of WebLab in laboratory exercises.

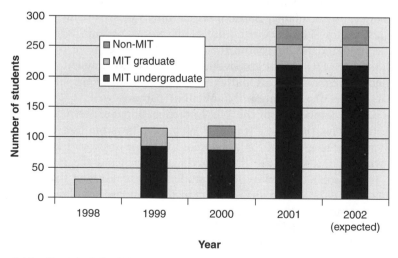

Figure 2.12   Chart depicting WebLab usage since 1998. In 2001, a total of 285 students used WebLab in class assignments. By the end of 2002, cumulative number of WebLab users is expected to exceed 800.

2.4.2.1 *Use of WebLab in Graduate and Undergraduate Subjects at MIT*   We have developed two kinds of homework assignments based on Web-Lab. A typical *WebLab homework problem* asks students to take some specific measurements on a microelectronics device, to graph the data in some standard format, and to download and further analyze the data using some other software. Typical devices are *pn* diodes, Schottky diodes, *n*-channel and *p*-channel MOSFETs, and *npn* bipolar transistors. In a typical homework problem, students are asked to obtain a standard set of characteristics of a particular device, to graph them properly, and to extract some relevant device parameters. The same homework commonly includes other numerical or analytical problems around the same device type. A standard WebLab homework problem can be carried out in 1–2 hr.

A *device characterization project* is a much more demanding and involved assignment that takes 10 or 15 hr and includes several elements:

1. Measurement of several types of device characteristics on a given device (for a MOSFET in a graduate class, they would be output, transfer, backgate, and subthreshold characteristics).
2. Graphing of the obtained characteristics in prescribed ways (using the applet or after data download).
3. Downloading the data onto the student's local machine.
4. Extraction of multiple-device parameters from the data set (for a MOS-FET, examples are threshold voltage, body parameter, surface potential at threshold, *k*-factor, channel length modulation parameter, subthreshold slope, and off-current, among others).
5. Programming a model based on the equations that describe the device operation as presented in lecture (in the case of a MOSFET, the model will involve a set of equations that smoothly describe the *I–V* characteristics of the device in the linear and saturation regimes and involve finite output conductance).
6. "Playing back" of the model using the extracted parameters as inputs against the obtained data.
7. Measurement of other device characteristics of the student's choice.

Figure 2.13 shows a representative set of results from an exercise like this involving an *n*-channel MOSFET where the measured and modeled characteristics of the device are plotted together using a standard spreadsheet program.

Item 7 in the list above, open-ended exploration, embodies the engineering laboratory experience at its best and is made possible by the flexibility of WebLab's measurement specification frame. In response to a request of this kind, students typically study other device characteristics that are not emphasized in class but that they have seen in other books. Some examples for a MOSFET are the diode characteristics of the source-body and drain-body

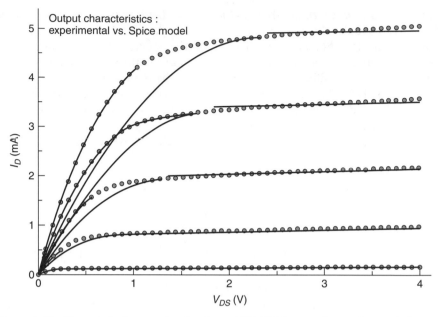

Figure 2.13 Characterization results of *n*-channel MOSFET mapped over characteristics predicted by theoretical model. Chart was plotted in standard spreadsheet program with data obtained and downloaded from WebLab.

pn junctions, body current characteristics under regular device operation, and observation of mobility degradation due to vertical electric field.

Device characterization projects constructed around WebLab have now become the norm in MIT's graduate subject on microelectronics device physics (6.720). In this class, three assignments are given every year (*pn* diode, MOSFET, and bipolar transistor). Simpler device characterization projects involving more elementary models are now also common in MIT's undergraduate subject on microelectronics devices and circuits (6.012). Since the spring of 2001, all offerings of 6.012 have included WebLab exercises.

2.4.2.2 Use of WebLab in Singapore–MIT Alliance In the fall of 2000, WebLab was deployed for the first time in SMA5104, a subject offered by the Singapore–MIT Alliance (SMA). The SMA was established in November 1998 between the National University of Singapore (NUS), Nanyang Technological University (NTU), and MIT to promote global engineering research and education. The SMA offers several graduate degree programs in various areas of engineering. The SMA relies heavily on advanced distance-learning facilities to transmit real-time educational content from MIT to Singapore.

The Advanced Materials for Micro- and Nano-Systems (AMMNS) program within the SMA exposes students to the foundations of processing, microstructure, and properties and performance of advanced materials, with particular emphasis on microelectronics applications. It is in SMA5104, "Fundamentals of Semiconductor Device Physics," a course of the AMMNS program, where WebLab was deployed to about 20 students in the fall of 2000 and to about 30 students in the fall of 2001. An interesting feature of this experiment is that the lab and the instructor and teaching assistant were located at MIT, while the students were located in Singapore. As in 6.720, the SMA5104 WebLab assignments required extensive characterization of an *n*-channel MOSFET, device parameter extraction, and device model generation and playback.

The feedback received from the students in Singapore was similar to that obtained from the MIT students that used WebLab in 6.012 and 6.720: mostly excitement with the novelty of the remote laboratory experience combined with expressions of minor frustration with a few aspects of the system. Some suggestions for improvement were given by the students. Most of the suggestions were related to the user interface and some reported bugs that we had not identified previously. No specific issues associated with the 9000 miles that separate MIT and Singapore were identified.

2.4.2.3 Use of WebLab from Compaq   In a third experiment, carried out in the fall of 2000, we installed a copy of WebLab at Compaq's Alpha Development Group center in Shrewsbury, Massachusetts. The purpose of this was to allow students in MIT's microelectronics graduate subject 6.720 to have access to the then state-of-the-art 0.18-µm CMOS device technology that Compaq's engineers were using to design leading-edge microprocessor products. In the fall of 2000, silicon wafers belonging to the 0.18-µm technology generation were closely guarded by microelectronics companies. Having a wafer on campus for our students to characterize was clearly out of the question due to intellectual property concerns. These concerns were eliminated by installing a copy of the WebLab system at Compaq's site and having the students characterize them remotely.

The single-device version of WebLab was installed around an Agilent 4155B that belonged to Compaq. The devices were not packaged; whole wafers were mounted on a standard wafer prober and single devices were probed. To avoid corporate firewall problems, the system was made accessible through a phone line. Through this system, MIT's graduate students were able to take remote measurements of 0.18-µm CMOS hardware, download the data, and compare the performance of this technology with 10-year-old 1.5-µm hardware that was made available through the on-campus WebLab system. For the first time, 6.720 students were able to study short-channel effects and other interesting device physics phenomena of modern deep-submicrometer CMOS technology.

This experiment showed the flexibility of the remote laboratory concept, which in several ways goes beyond what is possible in the traditional engineering laboratory. In a conventional "bricks-and-mortar" hands-on laboratory, it would be hard to expose students to leading-edge semiconductor technology because few corporations would be willing to lend the wafers. Packaged devices would be a suitable compromise if the corporate partner were willing to entertain the effort to die, package, and document the chips, but that would introduce significant extra work and reduce the partner's willingness to participate. By providing direct access to the technology at the corporate partner's site, intellectual property concerns are erased, and a novel and rich educational experience is enabled.

2.4.3 Summary of Lessons Learned

In the four years that we have been using WebLab in education, we have learned a number of lessons. Some of them are specific to the microelectronics device characterization experience that WebLab captures, but others can be generalized to remote laboratories in engineering education. A summary of the lessons that we have learned follows. The opinions of the students were polled through written questionnaires and interviews.

The first conclusion is that students are intrigued and motivated by the novelty of WebLab and, as a consequence, they pay considerable attention to WebLab assignments. Presumably, this results in more effective learning. To formalize this observation, we studied the homework assignments of 120 (mainly undergraduate) students enrolled in 6.012 at MIT in the spring of 2001. In that semester, there were seven regular problem-based homework assignments and three device characterization projects using WebLab (as described in Section 2.4.2.1). We found that the average score in the WebLab exercises was 83, while the average score in the problem set homeworks was 72. Also, 96% of the 6.012 students turned in solutions to the WebLab exercises, while only 92% did the same for the traditional homework. These statistics show that the nature of the WebLab experience catches the student's interests. This can be exploited to further the educational goals of the subject in which WebLab is inserted.

The second conclusion is that students dread real laboratories and appreciate the convenience of the remote laboratory experience. Engineering students at MIT, as in other engineering programs, have to take a number of laboratory-based courses. Students know what real hands-on laboratory experiences feel like, and they generally dislike them. In contrast, WebLab offers a new level of convenience that students really value. We have consistently heard three aspects of the WebLab experience that students greatly appreciate.

First, students tend to work late at night when it is unpleasant to get to and from the laboratory, particularly in inclement weather. With WebLab, they can do lab experiments from home.

Second, students tell us that using the hardware in most laboratories is rather frustrating. The tools are too complex, they do not work properly, or they have "quirks" that take time to figure out. Also, in electronics experiments, the cabling and the connectors often do not work well, and the wiring can be confusing. The simplified WebLab interface that presents only those controls that are needed to carry out the experiments minimizes opportunities for extraneous flaws that detract from the educational experience.

Third, the setup functions that are included in WebLab (see Section 2.3.1) are appreciated by students because these features allow them to work in a "stop-and-go" mode in which they can interrupt the lab experience at any time, save their work, and continue later at a convenient time. In a physical laboratory, this is rarely possible.

Students tell us that the convenience of WebLab when compared with the conventional laboratories allows them to focus on the educational issues. They report that they enjoy the lab experience more and learn better.

A word of caution and a disclaimer are appropriate here. The caution regards the risk of concluding from the lessons that we have learned that the engineering laboratory experience is best offered remotely through the Web. In fact, the authors disagree with this conclusion and in no way advocate the elimination of the traditional laboratory experience in engineering education. On the contrary, it is our belief that we should offer more opportunities for engineering students to carry out experiments on real systems, to face the difficulties associated with constructing and debugging complex experimental setups, and to deal with the issues brought along by real data. It is the logistical difficulties associated with delivering the traditional kind of laboratory experiences that has resulted in their diminished presence in the engineering curriculum. Through systems like WebLab, this trend can be largely redressed. While web-enabled remote laboratories do not capture the full richness of an actual hands-on laboratory experience, if properly constructed, they can go a long way in preserving their educational value and, in some cases, even enhance it.

The third significant finding from our WebLab experiments is that students, particularly in the undergraduate programs, have a great deal of trouble handling "real-world data." In retrospect, this is not surprising since there are not many experiences in the traditional engineering curriculum that expose students to this. We have learned that students often cannot distinguish "good" data (i.e., well-behaved transistor characteristics) from "bad" data (e.g., noisy output coming from a faulty experiment or the characteristics of a damaged transistor). They also have great difficulty manipulating measured data, such as graphing the data in another program or extracting parameters that describe a real device. We have also learned that students have very little intuition about what it takes to compare measured data with the theoretical models presented in class. Often students find themselves at a loss when they realize that the data never fit the model perfectly.

The fourth lesson we have learned is that it is difficult to integrate the WebLab assignments (mainly the device characterization projects) with the rest of a subject. This is probably the case because none of the subjects in which we have deployed WebLab had ever had a lab component in the past. In order to take advantage of WebLab, we have found that a certain degree of subject redesign must be carried out.

We have experimented with the location of the WebLab exercises in the course flow relative to the formal presentation of the underlying theoretical material in lecture. Placing the device characterization projects *after* the device in question has been presented in lecture is the most natural approach. However, in our microelectronics subjects, the most interesting devices are presented very late in the subject, and there might not be enough time for a substantial device characterization experience. Further, our subjects typically include a substantial design project toward the end of the course. This prevents the insertion of a weighty device characterization experience near that period.

To mitigate these difficulties, we have been experimenting for a few semesters with the placement of the device characterization projects *before* the relevant material is presented in lecture. We have found an unexpected advantage to this approach—beyond a greater degree of flexibility in the subject design—which is that students are significantly motivated by the WebLab assignments and give a heightened degree of attention when the relevant material is presented in lecture. This is seen in the increased quality and quantity of the questions that students ask in lecture. Still, the drawbacks of this approach are also apparent, mainly that the assignments can easily take a "cookbook" nature where the students follow recipes without really understanding what they are doing or why they are doing it. This can be alleviated to some extent by introducing in the assignments suitable explanatory material that summarizes microelectronics device behavior.

In general, we have found that WebLab exercises are not easy to design. This is probably not different from the content of laboratory subjects in general. However, inserting WebLab into a subject that did not have a laboratory experience in the past represents a substantial challenge. We have found that it is difficult to appreciate how much effort it takes for students to carry out the assignments and where their difficulties lie. In particular, as mentioned above, we have realized that students have little intuition about data manipulation, so seemingly simple tasks to an experienced engineer turn out to require substantial hand holding for an undergraduate student.

2.5   COLLABORATION

Student collaboration is a key component in engineering education; however, WebLab was designed for each student to work alone. The lack of collaborative features in WebLab restricts its potential significantly. Using the current WebLab design, one of two situations occurs. In one, students work in isolation. This removes the elements of collective analysis and group discussion from the students' lab experience. In the other, students who wish to collaborate physically gather around one computer terminal to use WebLab. Such a scenario is also undesirable because it places geographic requirements on students who wish to collaborate. We have recently developed a collaborative prototype that allows users in different locations to jointly participate in a WebLab session. Our experience is described in this section.

Exactly what features enable remote collaboration? Mark et al. (1999) dissected collaborative technology into two main categories: *communication tech-*

nologies and *information-sharing technologies*. Communication technologies let people who are collaborating discuss their work with each other. Chat, email, or conference calls are examples of this category. Information-sharing technologies are those that give one the ability to make other users aware of his or her activities and the results of those activities. File-sharing technology falls under this category.

Communication technologies abound in today's society, ranging from simple email and instant messaging to the more complex audio and video conferencing. In contrast, information-sharing technologies are much more difficult to implement, since they require the well-coordinated transfer of complex data. As we attempt to build a collaborative WebLab, we have focused our efforts on enabling the effective sharing of WebLab-specific information in real time.

We have implemented a prototype for a collaborative WebLab using Java servlet technology. This Java servlet prototype, hereafter referred to as the JS prototype, enables the sharing of WebLab inputs and outputs among a group of WebLab users without confiscating the individual user's control over the experiment setup. The JS prototype is not yet a complete application, but it encapsulates our proposed architecture for a collaborative WebLab.

2.5.1   Design Goals for a Collaborative Prototype

The Java applet is the only component of WebLab with which the user interacts. From the user's perspective, the interaction with the applet involves the input and output of a variety of data. Through the `Measurement Specification` frame, the user enters numerical parameters, defines control functions, selects measurement modes, and performs other miscellaneous tasks. Through the `Measurement Results` frame, the user receives graphical and textual presentations of data points returned from the server. Most of these inputs and outputs are difficult to describe through conventional means of communication, such as text chat or telephony. However, in a collaborative setting, some of this information needs to be shared between the collaborating users. The purpose of our prototype is to achieve the sharing of information that would enable efficient collaboration.

The design goals of the JS prototype are, therefore, as follows:

- *Enable Effective Sharing of WebLab-Related Information between Users*. The users who collaborate on WebLab should have the ability to share enough information with each other that they can easily become aware of their partners' progress.

- *Provide an Environment in which WebLab Users Can Collaborate with Minimal Conflict and Confusion*. In a virtual collaboration setting, clear communication is more difficult than in a physical collaboration setting. Confusion and conflict become harder to manage. We must minimize them to avoid wasting collaboration time in confusion clarification and conflict resolution.

- *Maximize Accessibility of the Application.* The main benefit of WebLab lies in its ability to bring the lab equipment outside of the physical laboratory. Wide accessibility was, therefore, a major goal of WebLab itself. Collaborative WebLab must be accessible to any user that can currently access the WebLab system.
- *Minimize Impact Collaborative Features Have on User's Ability to Carry Out WebLab Experiments.* The collaborative features are an extension to the existing WebLab system. All the tasks that individual users can perform on the existing system must still be feasible in collaborative WebLab to both the collaborating user and the individual user.

The following section describes an architecture for a collaborative WebLab that accomplishes these goals.

2.5.2   Architecture of JS Prototype

The architecture of the collaborative JS prototype is shown in Figure 2.14. Three components were added to the WebLab architecture to achieve our design of the JS prototype. A Java servlet was implemented as a *collaboration server*. A new, *dual-domain* user interface was introduced to handle collaboration. Our design also features a *modification token* scheme to regulate the collaborative activities among users so that conflicts within groups can be avoided. This section describes each of these components.

2.5.2.1 Collaboration Server   Java servlet technology was chosen as the development platform for the collaboration server because it is based on Java, the same language in which the WebLab user interface was developed (del Alamo et al., 2002a). Using the same language in the collaboration server minimizes the amount of modification required on the existing applet. This agrees with our design goal of minimizing the impact the collaborative features have on the WebLab user experience.

The collaboration server resides on the WebLab server along with the other software components of WebLab (see Figure 2.14). It performs the important tasks of connecting WebLab users and routing data among them. Once initialized, the collaboration server constantly listens for incoming connection requests from WebLab users. WebLab users who wish to collaborate with other users can select an option on their applet that establishes a socket connection between the applet and the collaboration server.

Unlike a connection established under the HTTP protocol where the server automatically closes the connection as soon as it responds to the client request, this socket connection is *persistent* and does not close until specifically requested by the client applet or the collaboration server. This needs to be the case because the collaboration server must be able to push data to its clients at unannounced times. Without a persistent connection between the client and the server, the client would have to poll the server repeatedly in order to receive

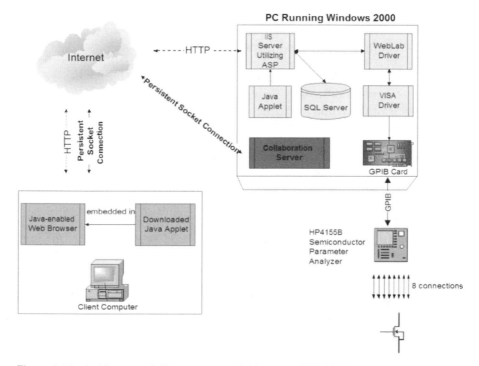

Figure 2.14 Architecture of JS prototype overlaid on top of WebLab's architecture (note that JS prototype was developed around single-device architecture). In our approach, *collaboration server* is only an extension to the existing WebLab system. Through persistent socket connections to collaboration server, users can share inputs and outputs.

updates from it. The repeated polling scheme places unnecessary burden on the client applet, and it does not allow the client to always receive data from the server in real time. A persistent socket connection is a much more effective solution for dynamic, bidirectional communication between client and server. When a client applet has established a persistent connection with the collaboration server, the server pushes to the client a list of all the currently connected users and notifies the other connected users of the new connected user.

The interaction between the client WebLab applet and the collaboration server demands a communication protocol. This protocol must be sufficient to achieve the sharing of all WebLab-related information necessary for collaboration. Yet, intensive communication between client and server places requirements on the quality of the network connection between them and thereby limits the accessibility of the application. To meet our design goal of maximizing accessibility, it is desirable to keep the interaction between the client applets and the collaboration server to a minimum. After examining all the inputs and outputs of the WebLab applet, we designed a text-based protocol with only 13 commands, including 7 client-side commands and 6 server-side commands (Chang, 2001; Chang and del Alamo, 2002). The applets can obtain the status of the other users, delegate the token, push data to the server, and pull data from the server using the commands in this protocol.

2.5.2.2 Dual-Domain User Interface

To achieve collaboration, the user interface of a collaborative groupware must provide to its individual users an awareness of the group's activities. However, in providing group awareness, the capabilities of the individual are often sacrificed (Gutwin and Greenberg, 1998). In an efficient collaborative setting, the individual should preserve his or her ability to work alone and contribute to or receive from the group only when appropriate. The JS prototype implements a dual-domain interface that offers significant group awareness yet gives the individual user a complete, private workspace. In this interface, each user maintains two domains—the *individual domain* and the *group domain*—each of which contains a complete WebLab interface. When the user first starts the JS prototype, only the individual domain is launched. This domain contains a complete WebLab interface, including an `Individual Measurement Specification` frame and an `Individual Measurement Results` frame. At this point, the applet is not connected to the collaboration server. As Figure 2.14 indicates, the collaboration server is an isolated extension to the WebLab system; so in the individual domain, the interface functions exactly as the single-user WebLab interface. Through this domain, the user can carry out individual experiments and execute any WebLab instruction.

When the user logs on to the collaboration server, three new frames appear: one frame that lists all the currently connected users and two frames of the *group* domain: `Group Measurement Specification` frame and `Group Measurement Results` frame. A typical desktop is shown in Figure 2.15. As can be seen in this figure, the group frames and the individual frames share nearly identical appearances. The two sets of frames have the same WebLab functionality, but the group frames are shared with the other connected users. The holder of the modification token (see Section 2.5.2.3 below) can modify the frames in the group domain. By means of the communication protocol, the token holder can also propagate his or her changes to the group domain of every connected user by selecting the `Synchronize` option. However, the individual domain remains under the control of the individual user—no other user has access to this domain. Thus, while the user could stay abreast of the group's collective progress by observing the frames in the group domain, he or she can also perform experiments in the individual domain without concern for unanticipated modifications by other users. Figure 2.15 shows the two domains of the interface side by side.

In the JS prototype, data can be transferred freely between the individual domain and the group domain. This enables a user to perform individual experiments based on the data in the group domain by *downloading* the data in the group domain into the individual domain. Alternatively, the user can share his or her individual work with the group by first obtaining the token and then

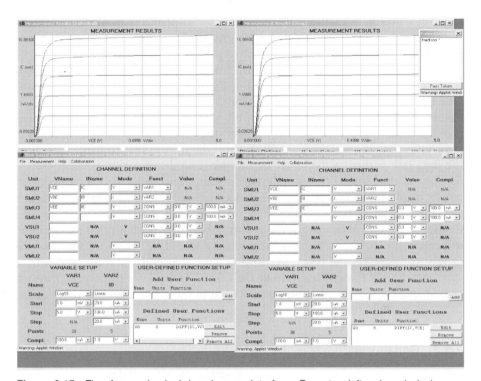

Figure 2.15 Five frames in dual-domain user interface. From top left going clockwise are `Individual Measurement Results` frame, `Group Measurement Results` frame, `Connected Users` frame, `Group Measurement Specification` frame, and `Individual Measurement Specification` frame.

uploading the data in the individual domain into the group domain. Once the data reaches the group domain, they are *pushed* to the other users.

2.5.2.3 Modification Token Scheme
Behind the user interface, there needs to be a set of logic regulating the group communication—rules that dictate who can communicate, when communication can occur, and what can be communicated. The key purpose behind communication control is the prevention of unwanted concurrent modifications of the group's objects (in the case of Web-Lab, the group's shared frames in the group domain). If multiple users attempt to modify the shared objects, they may overwrite each other's modifications and chaos inevitably occurs. In our prototype, we have introduced a modification token to ensure that concurrent modifications do not take place.

The modification token represents the right to modify the group's shared objects. Only one token exists in a group, and therefore, only one user holds the right to modify the shared objects at any moment. A user can obtain the token by default, by hand-over, or by inheritance. By default, the token is assigned to the first user who logs on to the group. The token holder can pass the token to any other user at anytime. If the token holder disconnects from the collaboration server before passing the token to anyone else, a randomly selected user inherits the token.

Under our token scheme, only the token holder can push measurement parameters and results to the server. For all other users, these controls are disabled. Any data received from the token holder by the collaboration server is automatically pushed to the applets of all other connected users, updating their group domains accordingly. If an applet does not hold the token, it can still pull data from the server. By pulling data, the applet synchronizes itself with the data last pushed to the server by the token holder.

This token scheme implies optimistic assumptions about the cooperativeness between collaborating users because it does not provide a method for the token to be taken away from a connected token holder. We assume that no user would "hog" the token against the group's objection. Without that assumption, a mechanism such as a group-voting scheme would be necessary to allow the group to "impeach the hog." We also assume that collaborating users would give each user a fair chance to hold the token. As we later discovered through tests of the prototype, our assumptions hold in the student groups that participated in our experiments.

2.5.3   Collaboration Experiments

In order to test the JS prototype, we recruited 21 MIT students for a preliminary set of experiments. We wished to investigate whether remote collaboration is useful from an educational perspective as well as receive feedback on the JS prototype's design. To minimize bias in the experiment results due to differences in the participants' backgrounds, all 21 students were enrolled in the same introductory microelectronics class at MIT that used WebLab for lab assignments.

The 21 participants were randomly categorized into three groups: the individual group, the collaboration group, and the tutorial group. We devised a moderately difficult, six-part WebLab exercise that mimicked those the students had encountered in their course assignments.

Six participants were assigned to the individual group. Participants in this group completed the exercises individually, without utilizing the collaboration features of the JS prototype. This is the reference group. Thirteen participants were assigned to the collaboration group. The members of this group were further split into two subgroups of two people and three subgroups of three people. Each subgroup had to complete the exercise as a group. The members of the subgroup worked in the same computer cluster, but they were prohibited from talking or gesturing to each other. They were required to collaborate using the JS prototype. A third-party chat program (either AOL Instant Messenger or Mirabilis ICQ) was also provided to them for message-based communication with each other.

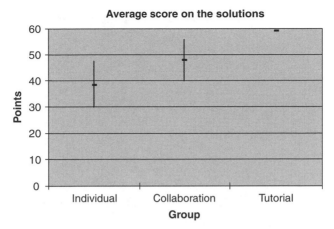

Figure 2.16   Average scores in WebLab exercise (each error bar represents two standard deviations).

Two participants were assigned to the tutorial group. Each participant in this group was given the JS prototype and a chat program just like the collaboration group. The participants completed the exercise individually but had access to an experienced tutor on the other end of the JS prototype when needed. The tutor was physically located away from the participant, so the setup simulated a remote tutoring environment.

During each experiment session, the data that flowed through the collaboration server and the chat conversation between the users were logged for later analysis. The completed WebLab exercises were collected and graded on an objective 60-point scale. We averaged the scores for each experiment group and compared the results (see Figure 2.16). As one would expect, the tutorial group, which had access to a tutor, performed the best. The individual group that proceeded without any collaboration scored the worst. These results, even though based on a small number of data points, suggest that remote collaboration improves student performance.

After each experiment in the collaboration and tutorial groups, the participants were given a survey that asked them for feedback on their experiences using the JS prototype. Participants were asked to rate the JS prototype in four categories on a scale of 1–7 (with 7 being the best rating): user friendliness, ease of using the token, ease of communicating with the group members, and ease of information sharing compared to face-to-face contact. The survey responses from the participants show promising results (see Figure 2.17). The average rating in each category is higher than 5. Even the question that asks the responder to compare collaboration in the JS prototype to face-to-face contact received a high average rating, which indicates that effectiveness of information sharing in the JS prototype approaches that of physical contact. At the end of

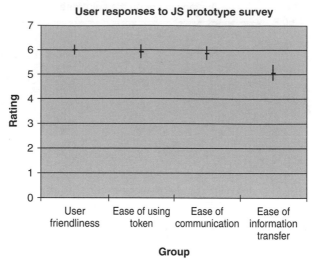

Figure 2.17 User feedback on WebLab experience involving JS prototype (each error bar represents two standard deviations).

the survey, the participants were given an opportunity to comment freely on the JS prototype. Of all the comments received, none of them were negative. The survey responses suggest to us that the JS prototype creates an effective collaborating environment for WebLab.

2.6 CONCLUDING REMARKS AND FUTURE PLANS

In 1998, we began the WebLab project with the purpose of using the Internet to make a state-of-the-art microelectronics device characterization laboratory affordable and accessible to students at MIT and beyond. The purely electronic nature of the instruments we intended to serve allowed us to deliver a rich lab experience over the World Wide Web. Through a Java applet that communicates with a server by means of HTTP, WebLab captures all the desired (and only the desired) functionalities of the lab instruments. We designed the system such that most tasks involved in measurement execution are performed off-line on the client machine. This keeps the lab instruments available and gives the system high scalability. The scalability of the system design and the widespread availability of Java-enabled web browsers enable the system to be efficiently shared among a large number of users in different locations. The goal of making the lab affordable and accessible to students has thus been attained.

Since its beginning, WebLab has been used routinely by undergraduate and graduate courses in microelectronics device physics at MIT and NUS. Previously, none of these courses offered any lab exposure, and WebLab has fulfilled their need for such a component. WebLab has served its purpose well, but we have repeatedly improved WebLab as we gained more experience with it. We learned that a single-device system is not sufficiently reliable and flexible for a large user base, so we created a multidevice system by adding a switching matrix to the WebLab architecture. We found that a remote laboratory needs a remote management system to be effectively administered, so we added a set of remote management features to the WebLab website. Lastly, realizing the value of collaboration in laboratories, we designed a suite of features to enable remote on-line collaboration in WebLab.

We still have plans to substantially advance WebLab in the future. Our future plans span three fronts: further development of the MIT Microelectronics WebLab, the exploration of a richer educational platform created around remote on-line laboratories, and the investigation of the extensibility of the WebLab concept to other engineering disciplines.

On the first front, we are considering many possible enhancements to our Microelectronics WebLab. Our plans include introducing a remote-controlled hot-chuck to vary device temperature, implementing a more intuitive graphical interface for test vector definition, and adding a new capacitance–voltage test capability.

We view remote on-line laboratories as the core of an educational platform with enormous educational value. This is what we call the I-Lab concept—a rich educational experience built around a web-enabled laboratory. Our second front of action focuses on pursuing that concept. Simulation tools can be integrated onto the WebLab platform so that one or several simulations can be launched simultaneously with the measurements. The results of simulations and measurements can be reported jointly through the same interface. We are also developing a collaboration/tutoring tool that will enable several remote users in separate locations to share the input and output frames of WebLab while "chatting" about the experience (Lin, 2002). Finally, "smart" semiautomatic feedback and evaluation systems can be added to assist users with common mistakes or to evaluate the degree to which a certain result complies with the assignment.

On the third front, together with a group of faculty at MIT, we are exploring the portability of the WebLab concept to other engineering disciplines. Several remote on-line laboratories are currently being developed at MIT. In the Department of Aeronautics and Astronautics, a mechanical structures laboratory is being constructed to study beam deformation in response to various stresses. In the Department of Chemical Engineering, a chemical reactor, a heat exchanger, and a microscope-based polymer recrystallization experiment are being designed and assembled. Finally, in the Department of Civil and Environmental Engineering, remote monitoring systems for a flagpole, a photovoltaic panel and a weather station, and model-building structures mounted on a shake table are under development. These laboratory experiments have been chosen to explore the WebLab concept across regimes where the physical variables of interest are of diverse natures and with time scale and length scales that span several orders of magnitude.

ACKNOWLEDGMENTS

The MIT Microelectronics WebLab has been developed through grants from Microsoft, the Microsoft–MIT Alliance (I-Campus), the Singapore–MIT Alliance, and MIT alumni funds (Classes of 1951, 1955, 1960, and 1972), as well as by equipment donations from Agilent Technologies, AMD, Hewlett-Packard, and Intel and software donations from Microsoft. We acknowledge the help of Ted Equi, Larry Bair, and Norm Leland for the experiments performed in the Fall of 2000 in collaboration with Compaq's Alpha Development Center and of Dimitri Antoniadis for the use of WebLab in the Singapore–MIT Alliance. We thank David Zych for a critical reading of this chapter.

REFERENCES

R. Berntzen, J. O. Strandman, T. A. Fjeldly, and M. S. Shur, "Advanced Solutions for Performing Real Experiments over the Internet," *Int. Conf. Eng. Ed. 2001*, p. 6B1-21 (2001). Available at http://nina.ecse.rpi.edu/shur/remote, 2001.

V. Chang, "Remote Collaboration in WebLab—an Online Laboratory," MIT Master of Engineering Thesis, MIT, Cambridge, Massachusetts, May 2001.

V. Chang and J. A. del Alamo, "Collaborative WebLab: Enabling Collaboration in an Online Microelectronics Device Characterization Laboratory," *Proceedings of Networked Learning 2002*, Paper No. 100029-03-VC-131, Berlin, May 1–4, 2002.

B. Dalton and K. Taylor, "Distributed Robotics over the Internet," *IEEE Robot. Automat. Mag.*, Vol. 7, No. 2, pp. 22–27 (2000). See also http://telerobot.mech.uwa.edu.au/.

J. A. del Alamo, L. Brooks, C. McLean, J. Hardison, G. Mishuris, V. Chang, and L. Hui, "The MIT Microelectronics WebLab: a Web-Enabled Remote Laboratory for Microelectronic Device Characterization," *Proceedings of Networked Learning 2002*, Paper No. 100029-03-JD-047, Berlin (Germany), May 1–4, 2002a.

J. A. del Alamo, J. Hardison, G. Mishuris, L. Brooks, C. McLean, V. Chang, and L. Hui, "Educational Experiments with an Online Microelectronics Characterization Laboratory," Paper No. 0102, presented at the International Conference on Engineering Education, Manchester, United Kingdom, August, 19–21, 2002b.

N. Ertugrul, "Towards Virtual Laboratories: A Survey of LabView-based Teaching/ Learning Tools and Future Trends," *Int. J. Eng. Ed.*, Vol. 16, No. 3 (2000).

C. Ferguson, "Remote Access to Computer-Controlled Manufacturing Facilities for Engineering Students," paper presented at World Manufacturing Congress, Auckland, New Zealand, 1997. On-going project at http://www.et.deakin.edu.au/research/Eng_Education/internet.htm.

C. Gutwin and S. Greenberg, "Design for Individuals, Design for Groups: Tradeoffs between Power and Workspace Awareness," in *Proceedings of CSCW '98*, New York, 1998, pp. 207–216.

J. Henry, "Running Laboratory Experiments via the World Wide Web," paper presented at the ASEE Annual Conference, Session 3513, 1996. Available from National Instruments Academic Resources CD-ROM 2001 Edition.

C. C. Ko, B. M. Chen, S. H. Chen, V. Ramakrishnan, R. Chen, S. Y. Hu, and Y.

Zhuang, "A Large Scale Web-Based Virtual Oscilloscope Laboratory Experiment," *IEE Eng. Sci. Ed. J.*, Vol. 9, No. 2, pp. 69–76 (April 2000).

C. C. Ko, B. M. Chen, S. Y. Hu, V. Ramakrishnan, C. D. Cheng, Y. Zhuang, and J. Chen, "A Web-Based Virtual Laboratory on a Frequency Modulation Experiment," *IEEE Trans. Syst. Man Cybernet., Pt. C: Applicat. Rev.*, Vol. 31, No. 3, pp. 295–303 (August 2001a). See also `http://vlab.ee.nus.edu.sg/vlab/index.html`.

C. C. Ko, B. M. Chen, J. Chen, Y. Zhuang, and K. C. Tan, "Development of a Web-Based Laboratory for Control Experiments on a Coupled Tank Apparatus," *IEEE Trans. Ed.*, Vol. 44, No. 1, pp. 76–86 (February 2001b).

Y. Lin, "A Collaboration System and a Graphical Interface for the MIT Microelectronics WebLab," Master of Engineering Thesis, MIT, Cambridge, Massachusetts, May 2002.

G. Mark, J. Grudin, and S. Poltrock, "Meeting at the Desktop: An Empirical Study of Virtually Collocated Teams," in *Proceedings of the Sixth European Conference on Computer Supported Collaborative Work*, Dordrecht, The Netherlands, 1999, pp. 99–118.

C. Salzmann, D. Gillet, and P. Huguenin, "Introduction to Real-time Control Using LabView with an Application to Distance Learning," *Int. J. Eng. Ed.*, Vol. 16, No. 3 (2000). See also `http://iawww.epfl.ch/Staff/Christophe.Salzmann/MS_HTML/ChMS.html`.

H. Shen, Z. Xu, B. Dalager, V. Kristiansen, O. Strom, M. Shur, T. Fjeldly, J. Q. Lü, and T. Ytterdal, "Conducting Laboratory Experiments over the Internet," *IEEE Trans. Ed.* (August 1999). See also `http://nina.ecse.rpi.edu/shur/remote/`.

M. Shur and A. Bhandari, "Access to an Instructional Control Laboratory Experiment through the World Wide Web," in *Proc. 1998 Amer. Contr. Conf.*, Philadelphia, Pennsylvania, 1998, pp. 1319–1325.

K. Taylor and J. P. Trevelyan, "Australia's Telerobot on the Web," paper presented at the Twenty-Sixth Symposium on Industrial Robotics, Singapore, October 1995, pp. 39–44. See also `http://telerobot.mech.uwa.edu.au/`.

3

INSTRUMENTATION
ON THE WEB

T. Zimmer, M. Billaud, D. Geoffroy, and Y. Danto
Université Bordeaux I, 33405 Talence Cedex, France

J. Martinez, F. Gomez, and I. Gonzalez
Universidad Autonoma de Madrid, 28049 Madrid, Spain

H. Effinger, W. Seifert, and A. Wiegand
Fachhochschule Münster, 48565 Steinfurt, Germany

3.1   INTRODUCTION

This chapter presents the setup of a remote electronics engineering lab on an European scale. It has been developed in the framework of the European Community Socrates program. The participating countries were France (University Bordeaux), Spain (University Autonoma of Madrid), and Germany (University of Applied Sciences of Münster). The aim was to allow the use of powerful instruments via the World Wide Web. A number of instruments, located anywhere in the world, set in a virtual lab and a remote training class can train with these instruments via a web page. This distance lab has been tested and evaluated in a real-world experience from students making a remote lab exercise.

Teaching and learning microelectronics are particularly challenging for both professors and students because of the relentless pace at which the field evolves.

Lab on the Web: Running Real Electronics Experiments via the Internet
Edited by Tor A. Fjeldly and Michael S. Shur
ISBN 0-471-41375-5   Copyright © 2003 John Wiley & Sons, Inc.

It is necessary, even essential, that teaching concepts follow, as quickly as possible, evolution in the laboratory and in industry. Test procedures involving such high-technology instruments as semiconductor parameter analyzers, *LCR* meters (inductance–capacitance–resistance meter), impedance analyzers, and network analyzers represent a fundamental part of the problem. The high price of these instruments creates a huge disadvantage for educational institutions, both because the instruments are too expensive to acquire solely for teaching and because they cannot be replaced regularly to keep pace with advanced technologies as they are introduced.

Remote measurements offer an answer to this challenge. In this approach, several laboratories band together, making their combined inventory of advanced instruments available to all authorized users. Furthermore, this approach makes new, more flexible pedagogical methods possible. When innovative techniques are used to offer lab exercises over the Internet, it is possible for students to decide individually about the time and the speed of progression in a course. This approach allows student-driven, step-by-step learning at each individual's own pace.

In recent years, communication and information technology (CIT) has progressed rapidly to meet expanding needs. These developments offer new opportunities for education.

Through the Socrates–Minerva program, the European Community has contributed substantially to the development of CIT-based tools for remote learning. From 1998 to 2000, the Microelectronics Laboratory IXL of the University of Bordeaux 1 (France) has developed the RETWINE project in partnership with the University of Applied Sciences of Münster (Germany) and the University Autonoma of Madrid (Spain).

RETWINE stands for Remote Worldwide Instrumentation Network. This tool permits the use of powerful instruments via the World Wide Web. The different network architectures and server solutions implemented in Bordeaux, Münster, and Madrid are discussed in the opening section of this chapter. The second part describes the mechanics of accessing the system and the corresponding tutorials that are currently available. The third section describes a lab exercise. The final section presents an evaluation by students who have conducted lab exercises by using RETWINE.

3.2 RETWINE ARCHITECTURE

This section describes the architecture of the remote instrumentation laboratory used in the RETWINE project, beginning with a short description of the overall system architecture common to all three sites. In addition, a section about the details of the architecture and implementation at each laboratory is included.

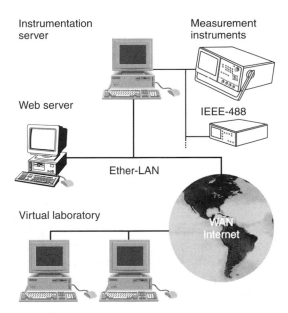

Figure 3.1 Instrumentation server scheme.

3.2.1 Overall System Architecture

The remote-access measurement system consists of (see Figure 3.1)

- the measurement instrument,
- an instrument server (a computer used to control the instruments),
- a web server, and
- client computers (personal computers or workstations used by students to access the system).

Most of the measurement instruments have a GPIB IEEE-488 interface card (general-purpose interface bus), allowing remote instrument control by an external computer. GPIB cards are available for all computer architectures and are usually delivered with libraries for high-level programming languages. Up to 30 instruments can be connected to the same GPIB card. Programming commands are sent by the computer as ASCII strings containing the target instrument number and instrument specific commands (e.g., RST, to reset the instrument to its initial state).

Remote measurement control involves three steps:

1. The stimuli and instrument options are set to the desired values (e.g., for a voltage source: values for integration time, number of measurement points, etc.).

2. A trigger command is sent to the instrument to start the measurement. The measured values are stored in the instrument's data buffer, and the instrument sends a "request to service" message indicating that the measurements have been completed.

3. Finally, the measurement data are transferred from the instrument to the computer.

On the local-area network (LAN), the instrument server acts as a TCP (transfer control protocol) server. It transmits information between the measurement instruments and the web server, which is connected to the Internet.

3.2.2 System Architecture Used in Madrid, Spain

3.2.2.1 Standard Architecture In Madrid, the controller program running on the instrument server is not tied to a particular type of measurement instrument or to a specific kind of analysis or test. It manages only the data exchange between the LAN and the GPIB interfaces, and it is able to serve several user requests for several instruments at the same time.

The computer acting as the web server fulfils various objectives (Cervera et al., 1999). First, it contains the web pages that allow users to connect to the remote measurement system. It also contains the necessary information about this remote system and the measurement instruments available through this server as well as information about instruments available from other servers in the RETWINE network.

As a second task, this server authenticates users and checks for reservations: To access the measurement equipment, the user must have a valid user account, which can be obtained by sending a request to the webmaster. Even with an account that allows full access to experiments, a user also must reserve a time slot. Booking guarantees that only one user has access to the instrument at any given time.

Finally, the server is a communications bridge for the information flow between the user and the accessed measurement system. As in the case of the instrument server, the information flow routing between the measurement instrument and the user does not depend on the measurement instrument. It is instead generic for any characterization or test that is being performed. The client–server software environment is shown in Figure 3.2.

On the client side, the web browser runs a platform-independent Java application (applet). Because communication between the user and the lab is an important factor in the success of remote measurements, audio and video communication facilities have been added. NetMeeting allows communication between the user and the operator in charge of the equipment. The experimental system can be monitored through video conferences.

3.2.2.2 An Improved Architecture Recently, new GPIB-LAN adapters have been developed (Tektronix, 1998). Such adapters have a GPIB interface

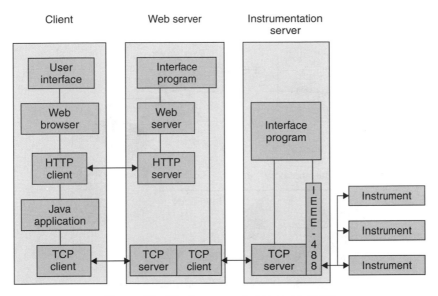

Figure 3.2   Client–server software environment.

and an Ethernet interface, and they implement all the server functions performed by the instrument server. The new architecture is shown in Figure 3.3. The GPIB-LAN adapter, which is less expensive than a computer with a GPIB card, supports the "VXI-11" specifications, which correspond to a network instrument protocol based on RPC (remote procedure call).

Although libraries from C++ on Windows are available, we have developed a set of Java classes in order to facilitate communications between the adapter and a web server that is using the RPC protocol. Several procedures have been defined for the GPIB-LAN server to perform all the steps needed to implement the measurement process. With this architecture, the web server performs booking and authentication functions (Gòmez et al., 2000). The new software environment is shown in Figure 3.4.

3.2.3 System Architecture Used in Bordeaux, France

Within the framework of the RETWINE project, the main objectives at the University of Bordeaux were to add the Hewlett-Packard HP4194 impedance/gain–phase analyzer and the HP8510 network analyzer. To accomplish this task, interfaces for these measurement instruments have been developed. These interfaces differ in programming details, but they comply with the same architecture (see Figure 3.5).

3.2.3.1 General Architecture   The web server (WS) provides tutorials and general information about the instruments as well as the Java applets that

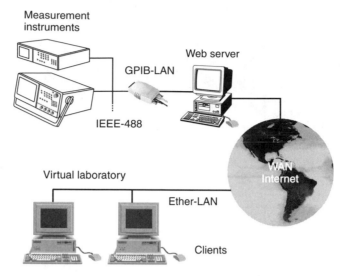

Figure 3.3 Improved server architecture scheme.

mimic the real instruments. These applets run on the client's web browser (Netscape Communicator or Microsoft Internet Explorer). They show the front panel of an instrument and transmit a request to the WS every time the user clicks on a button or a menu element.

The WS then asks the instrument server (IS) to transmit the request to the measurement instrument. This task is performed by a small driver program

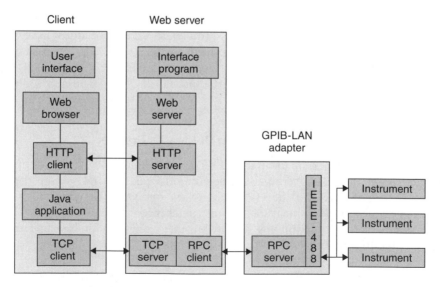

Figure 3.4 Client–server software environment with GPIB-LAN adapter.

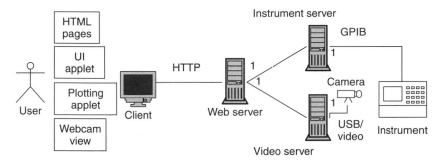

Figure 3.5 System architecture used in Bordeaux.

that talks to the instruments through the GPIB/IEEE-488 interface. The driver also stores measurement data in local files so they can be retrieved later.

In addition, the WS provides screenshots of the instruments: WebCams connected to a video server (VS) transmit live pictures of the instruments, and a video camera is attached to a framegrabber card.

3.2.3.2 How It Works The aim of the project is to teach students how to use real instruments. Consequently, the interface applet has been designed as a simple remote controller for the real instrument, displaying an image of the front panel that reacts to user mouse clicks in the same way that the real instrument does when keys are depressed on the front panel.

Unfortunately, the image of the instrument screen is far too small to display the measurement curves and menu panels. As a trade-off between convenience and picture realism, it was decided to draw the curves and display menus in separate windows. Also, numeric data can be entered using the numeric pad of the keyboard.

The active zones of the screen image (buttons, switches, menus) launch two types of actions: They send requests to the instrument and/or they change the internal state of the instrument (selection measurement modes, changing menu labels, turning LEDS on/off, etc.). The main difficulty in programming such applets is that some instruments (e.g., HP4194A) literally have hundreds of menu entries with context-dependent behavior, and the applet has to reproduce these as accurately as possible. As a result, much of the instrument interface complexity has to be simulated in the applet.

Former versions of the applets use `Socket()` connections to the instrument server to transmit actions. Unfortunately, this does not work when the client connection goes through a filtering firewall, which is a more and more common situation. In fact, the standard `Socket` connections try to establish a direct connection between the client navigator and the instrument server. Using `HTTPUrl` connections solve this problem. They permit actions via a proxy.

On the IS, the instrument driver is a simple C program that has two main functions:

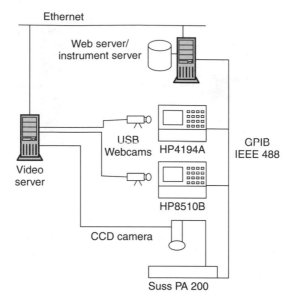

Figure 3.6 Remote laboratory at University of Bordeaux.

- It transmits requests to the instrument via the GPIB/IEEE-488 bus.
- In the case of measurement requests, it stores the results into measurement files so they can be retrieved later by the WS.

A small difficulty in writing these drivers results from the fact that some instruments do not report the end of measurement correctly, so various tricks such as inserting extra delays are used to avoid problems.

3.2.3.3 Implementation in Bordeaux

Currently the same computer (SUN SPARC Ultra 5/10 under Sun operating system 5.7) is used for the WS (Apache 1.3.12) and the IS (using a National Instrument IEEE-488 interface card). The remote lab implementation at the University of Bordeaux 1 is highlighted in Figure 3.6 (Billaud et al., 2002).

Student experiments are performed on an HP4194A impedance and gain/phase analyzer and an HP8510B network analyzer. Interfaces have also been developed for a Tek 11802B (digital sampling oscilloscope, Tektronix) and a Suss PA 200 Prober Bench (Suss Microtec).

The VS is a PC under Linux with a corresponding driver (ov511) for two inexpensive Webcam III USB (Creative Labs) and a Meteor II framegrabber card for the charge-coupled device (CCD) camera attached to the PA 200.

3.2.4 Software Architecture Used in Münster, Germany

In principle, the hardware setup and configuration used in the laboratory at the University of Applied Sciences in Münster is identical to the setup in the other two labs. The experimental equipment is connected via a GPIB to an interface card in a SPARC station running Solaris as an instrumentation server. All web-related information such as course material, instrument descriptions, and applets is accessible on a standard web server in the local university net.

To access the instrumentation, a user loads an applet from the WS with a standard web browser such as Netscape Communicator or Microsoft Internet Explorer. The applet uses standard TCP/IP to create a communication connection from the user's client computer via the web server to the instrumentation server.

3.2.4.1 Design Considerations

Two considerations were decisive in designing the software architecture. First, commands and data are exchanged between the applet on the client computer and the laboratory equipment connected to the instrumentation server. For this process, a protocol had to be defined. This protocol could have been either a connection-oriented or a connectionless protocol. In a first step, connection-oriented protocols like TCP in the IP suite must establish a connection between the two communication partners. Subsequently, commands and data can then be transferred over this two-point communication line until it is explicitly disconnected. Alternatively, in a connectionless protocol resembling UDP (User Datagram Protocol), all commands could be accumulated in the applet and sent to the device in one data packet without first establishing a standing connection.

The second important design consideration concerned the "intelligence" of the applet. As already mentioned above, the instruments of the remote lab may have many menu entries that allow changing the internal state of the instrument. If all commands generated by user interaction are first accumulated in the applet and are transferred to the instrument later in one data packet, the applet software has to duplicate much of the intelligence of the device. This would, of course, increase the complexity as well the size of the applet.

Since the applet code has to be downloaded by students, usually over low-bandwidth connections, we decided to use a lean-applet software architecture. The commands generated by a user interaction are immediately transferred to the device. Obviously, this implies a connection-oriented protocol for the exchange of commands and data between applet and device. But, of course, the use of lean applets and a connection-oriented protocol has the disadvantage that the device has to be reserved for one user and thus blocked for all others over a considerably longer period of time. Figure 3.7 shows the top-level data flow chart, that is, the context diagram of the software.

3.2.4.2 Software Architecture

The main modules of the resulting software architecture are sketched in Figure 3.8. The software comprises three components. The applet displays the front panel of the real instrument, the W-server component is executed on the server that houses the WS and the web pages, and the I-server is responsible for the interaction with the real devices over the GPIB. The bubbles in the diagram represent concurrent threads in the three components.

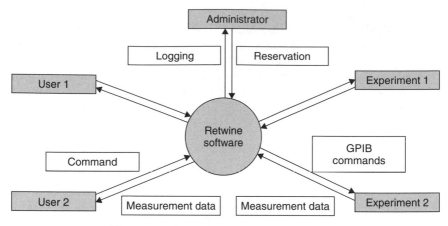

Figure 3.7 Context diagram of software design.

Figure 3.8 shows two different applets started by two users who are accessing different devices in the remote lab. The two bubbles, AR and AW, are separate threads in the applet. The applet writer, AW, is responsible for the transfer of commands from the applet to the device. The applet reader, AR, receives data from the device. One of the main reasons for using different threads for data transfer in the two directions is to avoid collisions during data exchange.

The central part of the communication software is the W-server program. It is the first process to be initiated during the start-up procedure of the virtual lab. Implemented in Java, the W-server is a multithreaded daemon running as a background process on the WS computer. Port reader threads, PR, within the daemon are started for all communication channels with incoming information,

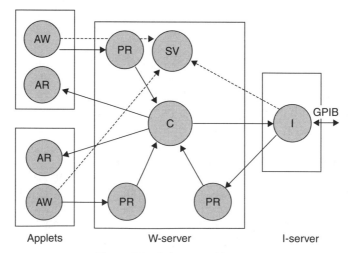

Figure 3.8 Software architecture.

Table 3.1   Example for Data Exchange Protocol

Source Address	Destination Address	Device	Command	Data
2 bytes	2 bytes	2 bytes	4 bytes	0–246 bytes

that is, for each applet and for the I-server module. The central communication thread, C, distributes all incoming data to the correct destination. In addition, it controls all global data structures in the W-server module. To guarantee data consistency and integrity, access to global data structures within the W-server is synchronized by appropriately implemented methods.

The task of the additional supervisor thread, SV, is administrative. The thread accepts connections to new clients, that is, applets or instrument servers. It maintains a list of active clients and starts a new port reader thread for each client.

Usually, the so-called I-server process is the second component started when booting the software for the remote lab. The program is implemented in the programming language C on the SPARC station connected to the GPIB. It is responsible for all activities related to the GPIB. In addition, the I-server checks user passwords and grants access to the remote lab.

3.2.4.3 *Protocols*   Several protocols have been designed and implemented for the exchange of commands and data between the different threads of the applet, the W-server and the I-server.

Table 3.1 shows an example of the protocol used for communication between the applet writer thread and the port reader of the W-server.

The complete specification of the data exchange protocol lies beyond the scope of this short description.

Summarizing, the software designed and implemented to grant access to the remote instrumentation pool at the University of Applied Sciences in Münster complies with a distributed three-tier model. The applet running on the client computer provides the graphical user interface for the devices, the W-server is the central communication module in the background, and the I-server manages the interaction with the real devices using the GPIB. The data exchange protocols between the components have been designed to provide a high degree of flexibility and extendibility. In a future extension of the functionality, for example, it will be quite easy to incorporate an applet-to-applet communication, thereby providing a direct connection between two users of the remote labs.

3.3   A WEB LABORATORY

3.3.1   Website: `www.retwine.net`

The homepage of the RETWINE project can be found at `http://www.retwine.net`. There is extensive information in the menu items of this

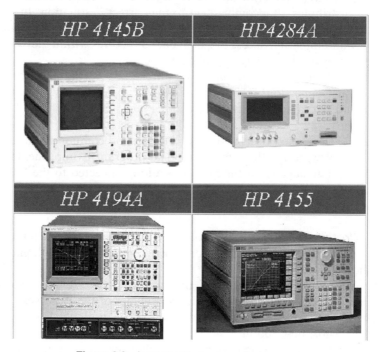

Figure 3.9 Instruments at www.retwine.net.

homepage concerning network architecture, server solutions, the partnership, and so on. To access the web laboratory, click on the Instruments button. A page displaying the front panels of a certain number of instruments appears on the screen, as can be seen in Figure 3.9. Select an instrument by clicking on the corresponding picture.

To access the measurement instruments, you need a valid user account, which can be obtained by sending a request to the RETWINE webmaster specifying your identity and your purpose for using the instruments.

3.3.2 Booking and Access to an Instrument

3.3.2.1 Booking an Instrument The University of Madrid has developed a booking system which guarantees that only one user has access to the instru-

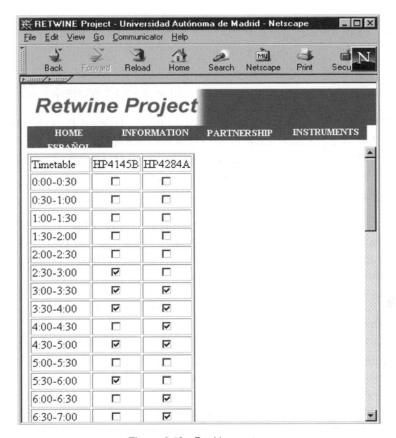

Figure 3.10   Booking system.

ment at a given time. This prevents others from using the instrument at the same time as you, which could lead to bogus results.

Once you have a valid user account, you can book the chosen instrument by using the RETWINE reservation system located at `http://micro.fa.uam.es/retwine/reservas`. First, you will be asked to enter your username, your password, and the day you want to book the instrument. Then the web page seen in Figure 3.10 will appear.

Figure 3.10 shows how booking is done. A checked box means that the instrument is already booked. You click on unchecked boxes to book the desired instrument for one of the time slots available. You then validate your choices by clicking on the `Book` button at the bottom of the page.

3.3.2.2   *How to Access Booked Instrument*   Each instrument (see Figure 3.9) has a menu item with some links. The link corresponding to the use of the instrument is `driver use`. Clicking on this item results in a dialog window that asks for username and password. After identity verification, the system makes sure that the instrument has been booked for the current time.

3.3.3   Tutorials

The following instruments are available within the RETWINE project:

- HP 4145B semiconductor parameter analyzer (Hewlett-Packard, 1984),
- HP 4284A precision *LCR* meter (Hewlett-Packard, 1988),
- HP 4155B semiconductor parameter analyzer (Hewlett-Packard, 1994),
- HP 8510B network analyzer (Hewlett-Packard, 1990), and
- HP 4194A impedance and gain/phase analyzer (Hewlett-Packard, 1991).

A tutorial has been created for each instrument, permitting students to learn the instrument's use step-by-step. The devices under test are standard components, such as capacitors, transistors, and filters. The various tutorials are listed below and are described in what follows:

1. Characterization of an *npn* Bipolar Transistor with the HP 4145B Semiconductor Parameter Analyzer.
2. $C(V)$ Characterization with the HP 4284A Precision *LCR* Meter.
3. Interactive measurement of *npn* Transistor (BF471) Characteristics with the HP 4155B Semiconductor Parameter Analyzer.
4. Transmission and Reflection Characteristic Measurements of a Filter with the HP 8510B Network Analyzer.
5. Measurement of a Capacitor's Frequency Characteristic with the HP4194A Impedance and Gain/Phase Analyzer.

3.3.3.1   Characterization of* npn *Transistor Using HP4145B   In this tutorial, the output characteristics of an *npn* transistor is measured. The collector current I_C versus collector emitter voltage V_{CE} with base current I_B as a parameter is displayed.

The transistor is connected to the test fixture HP 16058, as shown in Figure 3.11. The emitter E is connected to SMU1, the base B to SMU2, and the collector C to SMU3.

- To initiate measurements, push the on/off button on the instrument's front panel. The MENU page will appear as shown in Figure 3.12*a*.
- Press the Next button or the CHAN DEF softkey on the instrument's front panel to change to the CHANNEL DEFINITION page seen in Figure 3.12*b*. Fields must be filled or modified with the names used in the previous schematic. Depending on the field, the softkeys or the panel keys must be used. A message at the bottom of the screen helps the user to select the correct option. Figure 3.12*c* displays the desired selection: For SMU1, a

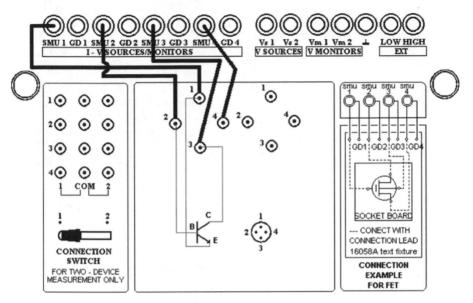

Figure 3.11   Device under test connected to HP4145B.

source mode COM is selected because $V_E = 0$ V. SMU2 is set as a current source I and assigned to the function VAR2, the secondary sweep. SMU3 is set as a voltage source V and assigned to the function VAR1, the main sweep.

- Press the Next button to change to the SOURCE SETUP page seen in Figure 3.13a. For VAR1, a LINEAR SWEEP mode is selected, varying V_{CE} from 0 to 5 V with a 0.05-V step. For VAR2, a LINEAR SWEEP mode is also selected; five steps of 10 μA are programmed, starting at $I_B = 10$ μA.

- Press the Next button to change to the MEAS & DISP MODE SETUP page, shown in Figure 3.13b. To show a graphic representation of the measurements, select the Graphics softkey. Move the cursor to the VCE-MAX field and change the value to 5 V.

- Press the Next button to change to the GRAPHICS PLOT page seen in Figure 3.13c.

- To start the measurement, press the Single button on the instrument's front panel. MEASURING LED (light-emitting diode) is on during the measurement. At the end of the process, the plot as seen in Figure 3.14 is displayed.

- Selecting the Autoscale softkey causes the graphic to be redrawn (Figure 3.15a) using an optimized scale with respect to measurement values.

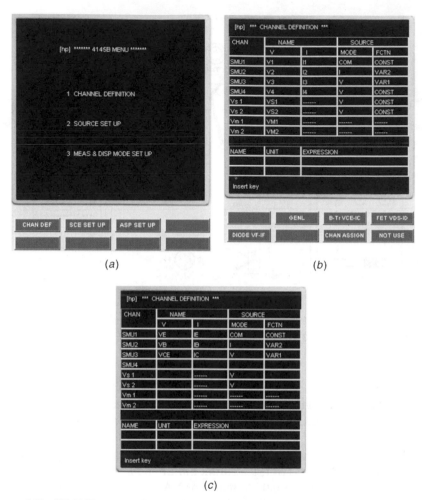

Figure 3.12 HP4145 menu and channel page: (a) menu page, (b) channel definition page, and (c) name assignment.

- When the `Marker` softkey is selected, a rhomboidal mark is shown at the x–y axes intercept. Use the mouse to drag the `Marker` dial in the front panel and rotate it. The marker will move along the plotted curves (Figure 3.15b). The x–y coordinates of the marker location will be displayed above the plot area.

- If the `Long Cursor` softkey is selected, a cursor is displayed in the graphic area (Figure 3.15c). The cursor can be moved with the mouse. Again, the x–y coordinates of the cursor location are displayed above the plot area.

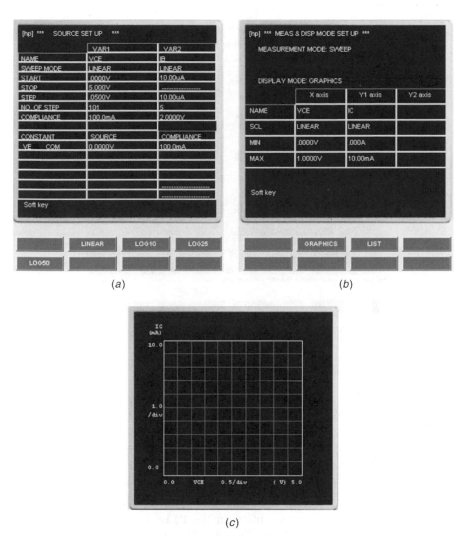

Figure 3.13 HP4145 source and measurement display setup: (*a*) source setup page, (*b*) display mode setup page, and (*c*) graphics plot page.

- In order to obtain a list of the measured data, press the `Prev` button to return to the `MEAS & DISP MODE SETUP` page. When the `List` softkey is selected, the `LIST DISPLAY SETUP` page is displayed (Figure 3.16*a*). Press the three available softkeys to select the monitor channels whose values will be listed. Press the `Next` button to change to the `LIST DISPLAY` page (Figure 3.16*b*). Select the `Single` button to start the measurement. At the end of the measurement, the values are listed. Only 10 lines can be dis-

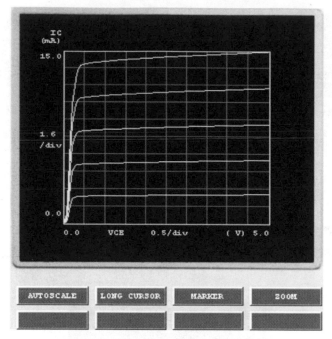

Figure 3.14 Graphical representation of measured data.

played. To display additional lines, use the `Roll Up` or `Roll Down` soft-keys.

- The measured data can be stored on the local computer by pressing the `Save` button. A new window will appear (Figure 3.17).
- Following the link `Click Here` from Figure 3.17 results in a new window (Figure 3.18) in which the measured data are listed in table form.
- Now select the `Save As` option in the `File` menu to store the measurement data to a local file. This ASCII file can be used by other programs.

3.3.3.2 C(V) Characterization Using HP4284A Precision LCR Meter

In this tutorial, the capacitance–voltage characteristic of a *pn* junction (a 1N4004 diode) is investigated. The capacitance is measured for five bias points on an HP4284A precision *LCR* meter. These include forward bias 0.2 V, reverse bias −1 V, and the intermediate bias points of 0, −0.2, and −0.5 V.

The diode has been connected to the equipment as shown in Figure 3.19, where the diode's anode has been connected to the `High` terminal. To prevent damage to the instrument or to the device, a high forward polarization should be avoided.

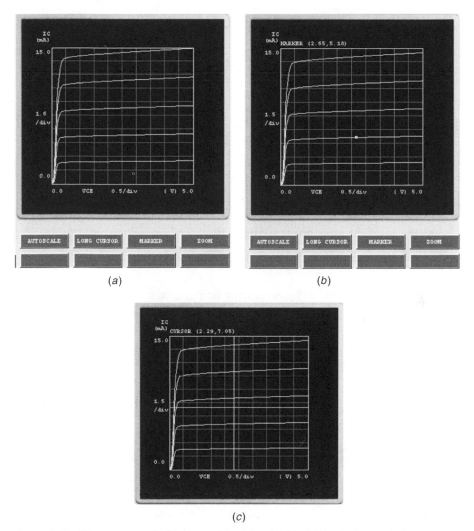

Figure 3.15 HP4145 curve display menu: (a) autoscale, (b) marker display, and (c) long cursor display.

Figure 3.20 shows the steps needed to configure the instrument:

Step 1: To initiate the measurements, press the On button. The instrument shows the screen <MEAS DISPLAY>.

Step 2: With the cursor keys, select the type of measurement to be performed. In the case illustrated, select Cp-G (capacity–conductance).

Step 3: Select the measurement frequency, which in this case is 100 kHz. Enter the value 100 with the numerical keyboard and press the kHz softkey.

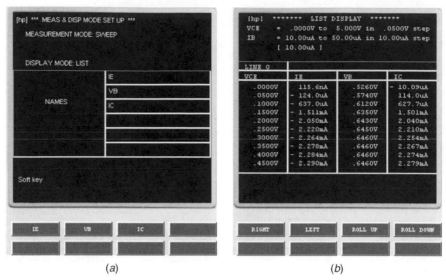

(a)
(b)

Figure 3.16 HP4145 list display menu: (a) list display page and (b) listed values.

Step 4: Select the measurement signal amplitude using a procedure similar to the frequency selection (keyboard-softkey). *Note:* The capacitance–voltage characteristic is nonlinear. To avoid measurement failure, the applied small signal amplitude should be small; select a value of 10 mV. The RANGE parameter must be set to its default AUTO value.

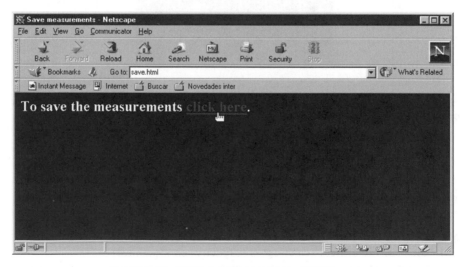

Figure 3.17 HTML link to save data in ASCII format.

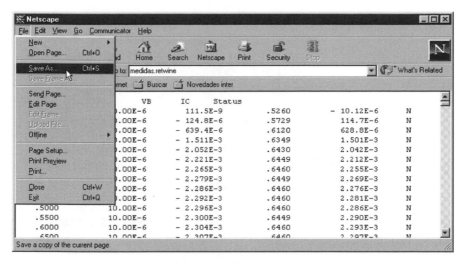

Figure 3.18 ASCII list exportable to other programs using copy and paste.

Step 5: Now it is necessary to select the bias voltage, beginning with a value of 0.2 V. On the screen, the measured capacitance value will be monitored.

The next action consists of repeating step 5 using different values of the bias voltage (Figure 3.21).

The capacitance values obtained for the different regions are summarized in Table 3.2.

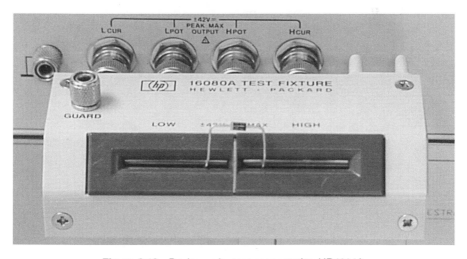

Figure 3.19 Device under test connected to HP4284A.

Step 1: Switch ON

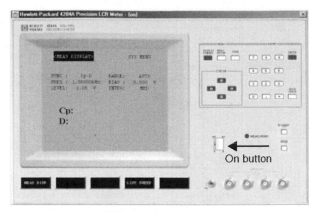

Step 2: Cp-G
Measurement Function

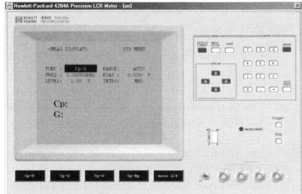

Step 3: Frequency
Selection

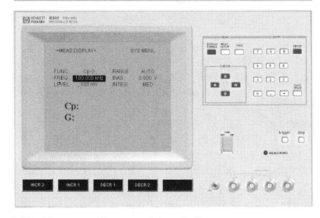

Figure 3.20 Measurement process (steps 1–3).

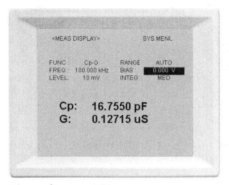

Bias voltage = 0 V

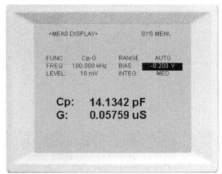

Bias voltage = -0.2 V

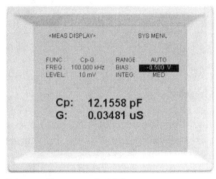

Bias voltage = -0.5 V

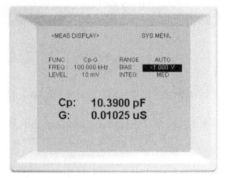

Bias voltage = -1.0 V

Figure 3.21   Measurement results.

3.3.3.3 Interactive Measurement Using HP4155B
In this section, measurements of semiconductor devices with the HP4155B semiconductor analyzer are described. If the HP4155B is connected via its respective I-server and W-server to the Internet, authorized users can perform measurements following the instructions on the following pages. The instructions lead step by step through the whole measurement procedure by demonstrating the operation of the HP 4155B. The device under test is a *npn* transistor BF471.

When you start a session using Internet Explorer, the windows seen in Figure 3.22 will appear on your screen: HP4155B, HP4155B-Panel, and Password Check.

Table 3.2   Experimental Data

Bias (V)	+0.2	0	−0.2	−0.5	−1.0
Capacitance (pF)	25.15	16.76	14.13	12.16	10.39

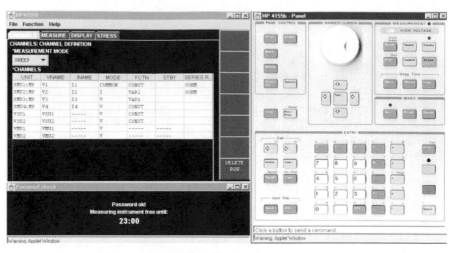

Figure 3.22   HP4155B windows.

Now, the following points have to be considered:

- If the `Password Check` window shows `Password ok!` you have access to the HP4155B until the time that is shown in the `Password Check` window.

- The results are not displayed using the `Graph/List` menu, but in a `Smart-Plot2D` window that appears on your PC screen once the measurement begins.

- It is not possible to access all the features of the HP4155B. Some features make no sense when used by remote control. Unlike the real HP4155B, the virtual HP4155B allows you to input using both your mouse as seen in the `HP4155B` window (=virtual screen, Figure 3.22, upper left window) and the keys seen in the `HP4155B-Panel` window (= virtual keypad, Figure 3.22, upper right window).

A very interesting supplementary feature is the temperature control of the device. The device is placed on a thermochuck, and the chuck temperature can be remote controlled. To access the temperature variation control of the DUT (device under test), it is necessary to select `Function` and then `Temperature` in the `HP4155B` window.

Before starting the *npn* transistor characteristics measurements, the device must be connected. Only three channels are needed. The measurement configuration is shown in Figure 3.23.

- For a standard default instrument configuration, reset the Hewlett-Packard parameter analyzer HP4155B. Next you have to select `Function` in the menu bar of the `HP4155B` window and then `Reset`. After that, the `HP4155B` window should be identical to Figure 3.24*a*.

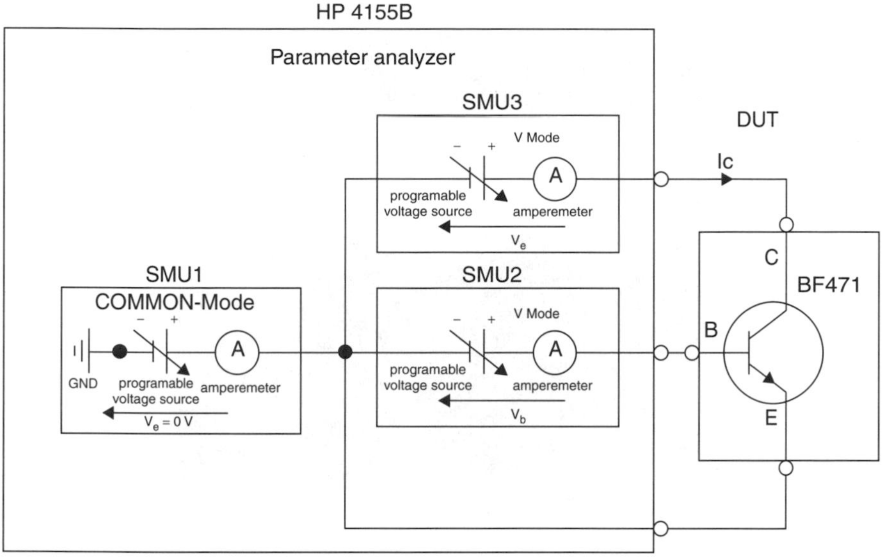

HP 4155B

Figure 3.23 Measurement configuration.

- You are now in the channel definition menu page. First, the measurement mode must be set to sweep: MEASUREMENT = SWEEP.
- To analyze a transistor, only three channels are needed, so channels SMU4 and VSU1–VSU4 can be deleted. To achieve this, select V4 in the second column (VNAME) and select the option DELETE ROW. After that, rows 5–8 must be deleted in the same way (see Figure 3.24*b*).

  Next you must configure the three remaining channels as follows:

  Channel 1 = SMU1:MP. As this channel is connected to the emitter of the BF471, the names and parameters are chosen as follows: VNAME = VE, INAME = IE, MODE = COMMON, and FCTN = CONST.

  Channel 2 = SMU2:MP. This channel is connected to the base of the transistor. To vary the current going into the base, the following names and parameters are chosen: VNAME = VB, INAME = IB, MODE = V, and FCTN = VAR2.

  Channel 3 = SMU3:MP. This channel is connected to the collector of the transistor. As the characteristics of a transistor are measured by varying the emitter–collector voltage, the following names and parameters are chosen: VNAME = VC, INAME = IC, MODE = V, and FCTN = VAR1.

- The option Series Resistance should remain at 0 Ω. The window HP4155B should now look like Figure 3.25*a*.
- By pressing the button MEAS in the PAGE CONTROL section of the HP4155B-Panel window, the Sweep Setup page is entered (Figure 3.25*b*):

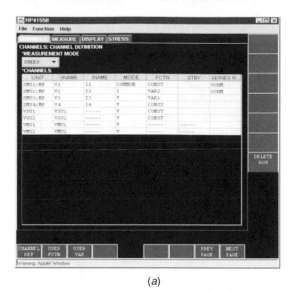

(a)

(b)

Figure 3.24 Channel assignment menu: (a) channel definition menu page and (b) selected values.

- Now, the range of the collector–emitter voltage must be set up. The following parameters are entered into the first column (VAR 1): SWEEP MODE = SINGLE, LIN/LOG = LINEAR, START = 0, STOP = 4, STEP = 0.05, and COMPLIANCE = 0.1.

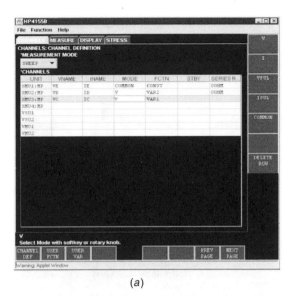

(a)

(b)

Figure 3.25 Stimuli menu: (a) final channel selection and (b) sweep setup page.

- Similarly, the base voltage is set up by entering the following parameters into the second column (VAR 2): SWEEP MODE = SINGLE, LIN/LOG = LINEAR, START = 0.6, STEP = 0.02, NO OF STEP = 4, and COMPLIANCE = 0.1.

- The *TIMING parameters HOLD TIME and DELAY TIME and the *CONSTANT and *SWEEP parameters are not changed, as shown in Figure 3.26a.

(a)

(b)

Figure 3.26   Sweep and display menu: (a) final values for sweep setup and (b) display mode.

- By pressing the DISPLAY key, which is located in the PAGE CONTROL section of the HP4155B-Panel window, the next setup page is entered. The HP4155B window should now look like Figure 3.26b. The outputs to be displayed in the Smartplot window are configured within this setup page.

Figure 3.27   SmartPlot2D dialog window.

- Now that the essential setup pages have been completed, the measurement can begin. This is accomplished by pressing the `Single` key, which is located in the `MEASUREMENT` section in the `HP4155B-Panel` window (see Figure 3.22, right side). Pushing the `Single` key will result in the display of the Smartplot window, signifying that the measurement has started (see Figure 3.27, left side). The end of the measurement is indicated by the message `Please Wait, Loading Data` (see Figure 3.27, right side).

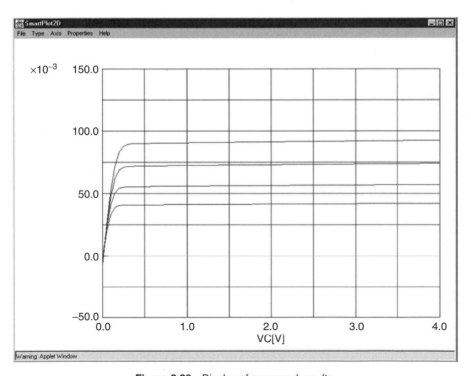

Figure 3.28   Display of measured results.

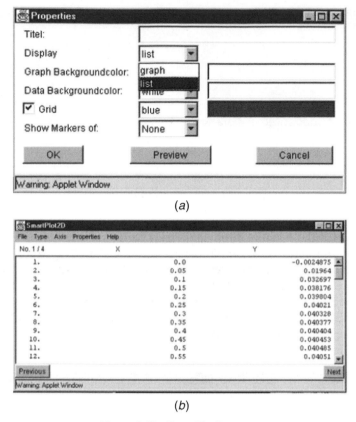

(a)

(b)

Figure 3.29 SmartPlot list menu.

- After that, the results are displayed graphically through the Smartplot window (Figure 3.28).
- It is also possible to retain a table containing each individual measurement instead of a graph. In the Smartplot menu click on `Properties` and then choose the option list in the `Display` menu (Figure 3.29a). You will get a list of the measured data as indicated on Figure 3.29b.

3.3.3.4 Use of HP8510 Network Analyzer This tutorial represents a step-by-step description of the use of the HP8510 network analyzer.

The first section describes the connection to the HP8510 via the Web. The second describes a typical measurement using this instrument, a measurement concerning the transmission and reflection characteristics of a filter.

3.3.3.4.1 Connection to HP8510

- First, connect to `http://retwine.ixl.u-bordeaux.fr:8080/retwine.html`.

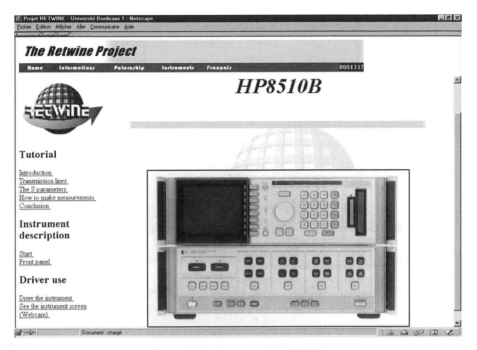

Figure 3.30   Start page for HP8510.

- Select the language: English or French.
- Click on `Instruments`.
- A page displaying the front panels of a certain number of instruments will appear on the screen. Select `HP8510B` by clicking on the corresponding image.
- Now your screen should look like Figure 3.30.
- On the left side of this figure, three parts can be distinguished: `Tutorial`, `Instrument Description`, and `Driver Use`
- In the tutorial section, you will find theoretical aspects such as an introduction to transmission line theory and to S parameters as well as some measurement examples.
- The instrument description section gives a detailed explanation of all of the instrument's hard and softkeys.
- Finally, the `Driver Use` section gives access to the instrument.
- To use the option `Drive the Instrument`, you need a username and a password, which you can obtain by contacting `retwine@ixl.u-bordeaux.fr`. Clicking on `Drive the Instrument` brings up the window in Figure 3.31.
- Enter your username and your password, and you will get access to the menu window shown in Figure 3.32.

Netscape: Password

Enter username for auth_hp8510b at retwine.ixl.u-bordeaux.fr:8080:

User ID: `Bienvenue`

Password:

| OK | Clear | Cancel |

Figure 3.31 Username and password dialog window.

- The first item in the HP8510 Driver window (right side) concerns the measurement driver. The next two items are related to the display of the measured curves, and the last permits you to download the measured data.

- Clicking on Start the driver leads to the screen display seen in Figure 3.33.

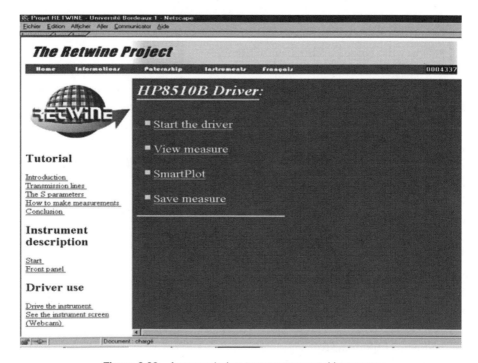

Figure 3.32 Access window to measurement driver page.

Figure 3.33   Front panel of HP8510.

- The next step consists of switching on the instrument by clicking on the switch in the lower left corner. You will now get two new windows, as shown in Figure 3.34.
- The left side of Figure 3.34 shows the softkeys. The labels of the softkeys depend on the instrument button pressed. In Figure 3.34, the default items are presented. The figure on the right side represents the front panel of the HP8510. The curve on the display indicates that the instrument has been switched on. Furthermore, the following message Ready should appear in the upper part of the window, indicating that the instrument is prepared for measurements.

3.3.3.4.2 Measurement of Filter Characteristics   After a user connects to the instrument as described above, measurements can be performed. The device to be tested must be connected to the instrument by the operator. In the following section, the transfer and reflection characteristics of a bandpass filter are measured. The results of the transfer characteristic are illustrated in Figure 3.35.

Figure 3.34   Front panel window and corresponding softkey window after switching on.

To perform the measurement, the following steps must be completed:

- Click on the Start button **Start** in the stimulus area of the front panel
  to define the start frequency of the frequency sweep. Enter a start frequency of 12 GHz by pushing the following sequence of buttons from the

  entry area of the front panel **1** **2** **G/n** .

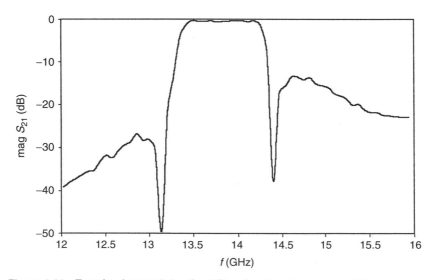

Figure 3.35   Transfer characteristics S_{21} (dB) as function of frequency of filter under test.

- Next, click on the `Stop` button 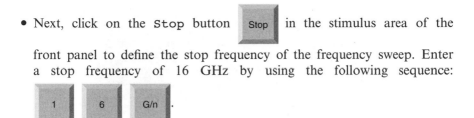 in the stimulus area of the front panel to define the stop frequency of the frequency sweep. Enter a stop frequency of 16 GHz by using the following sequence:

- In addition to these measurements, you can define the number of measured points, the power of the test signal, the applied sweep mode, and other characteristics. Refer to the description section of the web page for the HP8510 where all functions are described.

- Next, select the S21 parameter in the `Parameter` part of the front panel by clicking on `S21`. Then choose the magnitude by clicking on `Log mag` in the `Format` menu of the front panel.

- Clicking on the `Restart` button `Restart` located in the lower right corner of the front panel will cause the measurement to be performed.

- You can visualize the instrument screen by selecting `Webcam`. You will find it in the lower left section, as seen in Figure 3.32. Click on the `Autoscale` button `Auto` of the `Response` menu. In this way, the measured curves are optimally fitted to the instrument screen. The Webcam picture is illustrated in Figure 3.36.

- Finally, the measured data can be visualized directly by using the `View measure` or `SmartPlot` items (see Figure 3.32). It is also possible to download the measured data by clicking on the `Save measure` item. The data are available in ASCII format and can then be read by using normal spreadsheet programs such as MS Excel (Figure 3.35).

It is now easy to determine center frequency, cut-off frequency, and selectivity. Furthermore, the return loss of the filter can be determined and analyzed in a similar manner.

3.3.3.5 How to Use HP4194A Impedance/Gain–Phase Analyzer This tutorial describes how to use the HP4194A impedance/gain–phase analyzer. Use of the instrument is illustrated by a case study on the measurement of a capacitor's frequency characteristics.

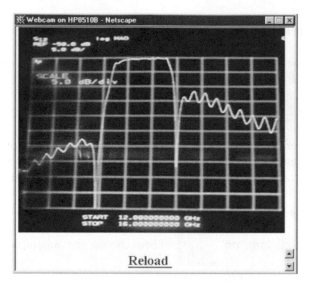

Figure 3.36 Webcam picture of instrument screen after measurement of filter transfer characteristic.

3.3.3.5.1 Generality First, connect to the HP4194 using a connection procedure similar to the one described in the HP8510 tutorial. After you are connected to the HP4194, two Java-type window pictures are created, and they appear on the screen. One shows the device front panel (Figure 3.37*a*). The other reproduces the HP4194A softkeys (Figure 3.37*b*), which inform the user about the communication status with the server and permit the user to enter the parameters needed to perform measurements.

3.3.3.5.2 Measurement of Frequency Characteristics of a Capacitor

- To begin, click on the button `Line` located under the screen of the front panel. The message at the top of the softkey window is `The HP 4194A is ready` (see Figure 3.38). The instrument can be used by simply pushing the desired keys. (*Note*: Some keys are not implemented, especially those that relate to the system configuration and to the markers.)
- Click on the `Function` button (indicated by arrow 1 of Figure 3.37*a*); a softkey window appears (Figure 3.38).
- To study the capacitor, choose the impedance of the HP4194A analyzer. Click on `Impedance` in the softkey window;
- Click on the `|z|-theta` mode in the following softkey window. You will measure the impedance and phase of the device.
- Notice the OK message in green letters at the top of the softkey window after each choice.

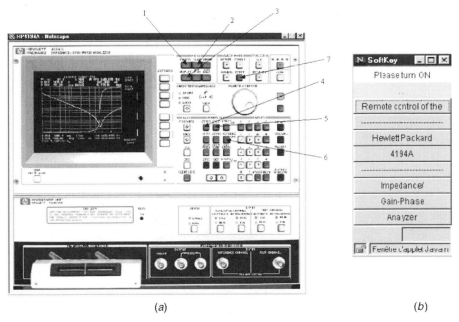

(a)   (b)

Figure 3.37 (a) HP4194A device front panel applet visualization and (b) HP4194 softkey window.

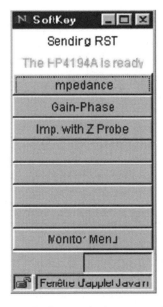

Figure 3.38 Communication status window.

Figure 3.39 Function softkeys.

- Click on the `Sweep` button (arrow 2). In the new softkey window, choose `log sweep` for a correct display of the Bode diagram.
- Click on the `Display` button (arrow 3) to access the display option. To choose a log or linear y scale on the display, click on `Menu` in the softkey window and then on your choice.
- Click on the `Start` button (arrow 4) and enter the desired value of the start frequency of the frequency sweep in the field located at the bottom of the softkey window (Figure 3.39) (100 Hz in the example).
- Repeat this operation with the `Stop` button (arrow 5) (400 MHz in the example). Remark: The units must be entered in capital letters and without a space.
- Click on the `N` button (arrow 6) to define the number of measurement points. `NOP =` appears in the text field of the softkeys window. You should only have to type a number (e.g., 50) and validate your entry by pushing the `Return` key of your keyboard. The string `NOP=50` is then sent to the HP4194A.
- Once these parameters are entered, you must start the measurement. Click on the `Start` button (arrow 7).
- You can visualize the instrument screen by selecting the Webcam item from the measurement driver page. Click on the `Display` button (arrow 3) and then on `Autoscale` in the softkey window. In this way, the curve is optimally fitted to the instrument screen. The Webcam picture is illustrated in Figure 3.40.

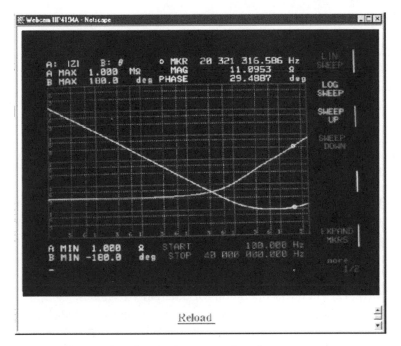

Figure 3.40 Graphical representation of measurements.

3.4 LABORATORY 1. ELECTRONICS OF INTEGRATED CIRCUITS: MOS TRANSISTOR

3.4.1 Aim of Laboratory Exercise

The following analog electronics lab exercise was designed for electrical engineering curriculums. As part of the RETWINE project, the exercise was completed and evaluated by the IUP students in electrical engineering (sixth semester) at the University of Bordeaux 1, France.

The objective is to familiarize students with basic MOS transistor characteristics and to extract from these characteristics fundamental technological and device data such as substrate doping density, oxide thickness, electron surface mobility, threshold voltage, body effect parameter, and output conductance (Zimmer and Geoffroy, 2000).

In a first step, the layout of a specific testchip is studied, and basic devices such as capacitors, diodes, resistors and transistors are identified. Next, $C(V)$ measurements are performed on a specific test structure, in which gate oxide thickness and substrate doping are extracted. Finally, a complete characterization of the MOS transistors is performed based on measured DC characteristics.

3.4.2 Keywords

- MOS transistor
- CMOS technology
- Device modeling

3.4.3 Prelaboratory Analysis

Before starting the lab exercise, the students are asked to complete the following prelab analysis:

(a) Draw the physical structure of an NMOS and PMOS transistor in a p-well, CMOS technology: top side view and cross-sectional view.

(b) Plot the electrical characteristics of a NMOS transistor:

$I_D(V_{DS})$ for different V_{GS} values, $V_{BS} = 0$;

$I_D(V_{GS})$ in saturation mode; and

$I_D(V_{GS})$ in linear operation for different V_{BS} values.

(c) Indicate how the potential of the substrate influences the electrical characteristics and to which nodes do the substrates (p-well, n-well) of the NMOS and PMOS transistor have to be connected, respectively.

(d) Give the definitions of weak inversion and strong inversion.

(e) Give the physical significance and the associated unit of each parameter in the following expression:

$$I_{D,\text{sat}} = \frac{\mu_0 C_{\text{ox}}}{2} \frac{W}{L} (V_{GS} - V_T)^2 (1 + \lambda V_{DS})$$

(f) Show how the expression for the transconductance g_m can be derived from $I_{D,\text{sat}}$. Give the expression of g_m as a function of $I_{D,\text{sat}}$ (in saturation).

(g) Indicate how the expression for the conductance g_{ds} can be derived from $I_{D,\text{sat}}$. Give the expression of g_{ds} as a function of $I_{D,\text{sat}}$ (in saturation).

(h) Show how the expression for the transconductance g_{mbs} can be derived from $I_{D,\text{sat}}$. Give the expression of g_{mbs} as a function of $I_{D,\text{sat}}$ (in saturation).

Reminder: The threshold voltage V_T depends on the bulk–source voltage V_{BS}: $V_T = V_{T0} + \gamma \lfloor \sqrt{2|\Phi_F| + V_{SB}} - \sqrt{2|\Phi_F|} \rfloor$, where V_{T0} corresponds to the threshold voltage when $V_{BS} - 0$ V, γ is the body effect parameter, and Φ_F is the surface potential at strong inversion.

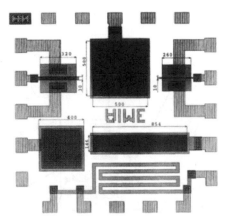

Figure 3.41   Chip layout.

3.4.4   Experiments

After completing the prelab work, students use the Web to access remote instruments to investigate MOS technology, study the layout of some test structures, measure some of these test structures, and extract the principal parameters to characterize the technology and the MOS device. Instructions to the students are as follows.

3.4.4.1 Study of Layout   The chip layout is presented in Figure 3.41. The bonding is indicated in Figure 3.42. With the use of the chip layout, identify the following devices:

- one long-channel NMOS transistor ($L = 10$ μm, $W = 260$ μm);
- one long-channel NMOS transistor ($L = 30$ μm, $W = 320$ μm);
- one MOS capacitance, area $= 500 \times 500$ μm;
- one MOS capacitance, area $= 146 \times 854$ μm;
- one n^+/p^- diode, area $= 400 \times 400$ μm; and
- one n^+ diffused resistance for «4 points» measurements ($L/W = 136$ squares).

Identify the different devices. Indicate the gate, drain, and source. The two capacities have the same gate periphery. Explain how this can be utilized.

3.4.4.2 Measurement of MOS Capacitance   MOS capacitance is one of the key test structures for MOS technology characterization. It permits the determination of the principal characteristics of a given technology such as oxide thickness and substrate doping. The test structure consists of a large gate area without any drain and source diffusion. The gate–bulk capacitance is

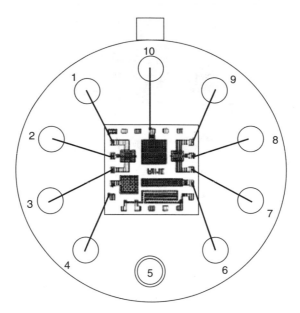

Figure 3.42 Chip bonding, top view.

measured as a function of a DC gate–bulk voltage. The capacitance measurement is performed by superimposing a small-signal voltage (amplitude of 30 mV, frequency of 1 MHz) on the DC bias. From the AC signal response of the device, the capacitance value can be determined. The corresponding calculations are directly executed by the measurement instrument.

Figure 3.43 gives a schematic cross section of the test structure as well as its corresponding equivalent circuit. Two series capacitances per unit area (C_{OX} and C_D) can be observed. Here C_{OX} represents the capacitance per unit area between the gate metal and the bulk surface. It has a constant, voltage-independent value, which depends on geometric properties, such as gate area and oxide thickness, and on material properties of the oxide, such as the dielectric constant ε_{SiO_2}. The parameter C_D represents the depletion capacitance per unit area and depends on the applied voltage.

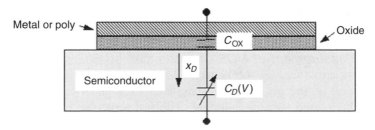

Figure 3.43 MOS capacitance test structure and its equivalent circuit.

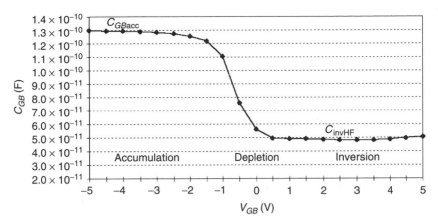

Figure 3.44 $C(V)$ behavior for MOS capacitance test structure measured at high frequency (1 MHz).

Figure 3.44 represents a typical $C(V)$ behavior for a MOS capacitance test structure measured at high frequency (1 MHz). The operation ranges are also indicated on this figure: strong inversion, depletion, and accumulation.

3.4.4.3 Determination of Oxide Thickness T_{OX}

Under accumulation ($V_{GB} \ll 0$ for a p substrate), holes (majority carriers) accumulate under the oxide–semiconductor interface because of the negative gate bias. The impedance related to the C_D capacitance can be neglected under this operating condition. All charges are located on the upper and lower interfaces of the oxide, and the measured C value corresponds to the oxide capacitance as follows:

$$C_{GB,\text{acc}} = \frac{\varepsilon_{OX} A}{T_{OX}} = C_{OX} A$$

where A is the capacitance area, $\varepsilon_{OX} = \varepsilon_{SiO_2} \varepsilon_0 = 3.45 \times 10^{-11}$ F/m the dielectric constant of SiO_2, and T_{OX} the oxide thickness.

Using the above expression, T_{OX} can be directly determined from the measured C_{GB} curve.

3.4.4.4 Determination of Substrate Doping NSUB

Under depletion and inversion, the charge on the gate is equilibrated on the semiconductor side by a space charge region under the gate.

The structure is also equivalent to two series capacitances, one due to the oxide capacitance and the other due to the capacitance of the space charge region. The measured capacitance value corresponds to the two capacitances in series, resulting in a smaller capacitance than that measured under accumulation.

Assuming constant doping for the substrate and using the measured C value C_{invHF} at strong inversion, NSUB can be deduced:

$$C_{invHF} = \frac{A}{1/C_{OX} + 1/C_D}$$

so

$$C_D = \frac{1}{A/C_{invHF} - 1/C_{OX}}$$

On the other hand, C_D is also given by Sze (1981):

$$C_D = \sqrt{\frac{q\varepsilon_{Si}\,\mathrm{NSUB}}{4\Phi_F}}$$

which results in

$$\mathrm{NSUB} = \frac{4C_D^2}{q\varepsilon_{Si}}\Phi_F$$

But it has also been taken into account that Φ_F depends on substrate doping NSUB:

$$\Phi_F = \frac{kT}{q}\ln\frac{\mathrm{NSUB}}{n_i}$$

where Φ_F is the surface potential at strong inversion, $n_i = 1.45 \times 10^{10}$ cm^{-3} is the intrinsic carrier density, $kT/q = 25.8$ mV is at $T = 300$ K, $\varepsilon_{Si} = \varepsilon_{Si}\varepsilon_0 = 1.04 \times 10^{-10}$ F/m is the dielectric constant of Si, and $q = 1.6 \times 10^{-19}$ C is the electron charge.

No analytical solution exists for NSUB, but an iterative numerical method permits the determination of NSUB. An initial choice of a NSUB value in the range of 10^{15}–10^{16} At/cm^{-3} permits a fast convergence to a stable and physical solution.

3.4.5 Measurement and Extraction

In this exercise, the student is able to measure the $C(V)$ behavior with the precision LCR meter HP4284. This instrument is located in Spain (Universidad Autónoma de Madrid). The measurement is performed as follows: On your personal computer, which must be connected to the Internet, start a browser program (Netscape or MS Internet Explorer). Load the URL http:// www.retwine.net, click on Devices, and choose the instrument HP4284. As a first step, consult the tutorials to learn how to use the instrument.

The device connected to the instrument in Madrid is a capacitance meter with the area $A = 250 \times 10^{-9}$ m^2. The capacitance meter is connected between the gate (pin 10) and the substrate (pin 5). Adjust the oscillation level to 30 mV and the oscillation frequency to 1 MHz and display C_p, R_p.

- Perform the measurements and vary the DC bias between ± 5 V.
- Record the measured values.
- Plot the characteristics $C(V)$.
- Determine C_{OX} and T_{OX}.
- Determine NSUB following steps (a)–(e):
  (a) Express n_i and NSUB in atoms per cubic meters.
  (b) Calculate α defined by the expression

$$\alpha = \frac{4C_D^2}{q\varepsilon_{Si}} \frac{kT}{q}$$

  (c) Choose a physical starting value for NSUB$_0$.
  (d) Calculate NSUB$_1$ applying the following expression:

$$\text{NSUB}_j = \alpha \ln \frac{\text{NSUB}_{j-1}}{n_i}$$

  (e) Repeat the previous step until NSUB$_j$ converge to a constant value (fill in Table 3.3).

3.4.6 Measurement of MOS Transistor DC Characteristics

Measurement of the MOS transistor DC characteristics can be completed as follows.

The equations describing the DC characteristics of the MOS device are of first order, neglecting short- and narrow-channel effects (Allen and Holberg, 1987):

$$i_D = \begin{cases} K'_S \dfrac{W}{2L}(V_{GS} - V_T)^2(1 + \lambda V_{DS}) & \text{saturation} & (3.1) \\[3ex] K'_L \dfrac{W}{L}\left[(V_{GS} - V_T)V_{DS} - \dfrac{V_{DS}^2}{2}\right](1 + \lambda V_{DS}) & \text{nonsaturation} & (3.2) \end{cases}$$

Table 3.3 Iterative Determination of Substrate Doping Density NSUB

j	0	1	2	3	4	5
NSUB (m^{-3})						

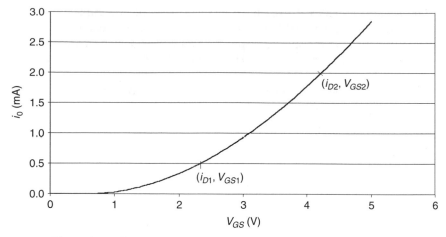

Figure 3.45 Drain current i_D as function of V_{GS} in saturation, $V_{DS} = 5$ V.

with

$$V_T = V_{T0} + \gamma(\sqrt{2|\Phi_F| + V_{SB}} - \sqrt{2|\Phi_F|}) \tag{3.3}$$

The first-order parameters are V_{T0} ($V_{SB} = 0$), K'_L, K'_S, γ, and λ.

3.4.6.1 *Extraction of Threshold Voltage V_{T0}*

In a first approximation, the channel length modulation parameter λ can be neglected. Under this assumption and with the bias condition $V_{SB} = 0$ ($V_T = V_{T0}$), Eq. 3.1 can be rewritten as

$$i_D = K'_S \frac{W}{2L}(V_{GS} - V_T)^2 \tag{3.4}$$

This is a parabolic function and is plotted in Figure 3.45.

Choosing two points, (i_{D1}, V_{GS1}) and (i_{D2}, V_{GS2}), on this graph, the parameters V_{T0} and K'_S can be determined. Additionally, when the two points fulfill the condition that $i_{D2} = 4i_{D1}$, it can be shown that $V_{T0} = 2V_{GS1} - V_{GS2}$. Then for K'_S it follows that

$$K'_S = \frac{2L}{W} \frac{i_{D1}}{(V_{GS1} - V_{T0})^2}$$

3.4.6.2 *Extraction of K'_L in Nonsaturation*

Under the assumption of a small drain–source voltage ($V_{DS} = 100$ mV), the channel length modulation

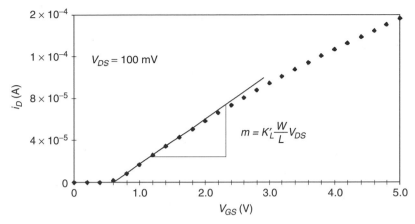

Figure 3.46 Drain current i_D as function of V_{GS} in nonsaturation.

parameter λ can be neglected. Equation 3.2 can then be rewritten as

$$i_D = K'_L \frac{W}{L} V_{GS} V_{DS} - K'_L \frac{W}{L} V_{DS} \left(V_T + \frac{V_{DS}}{2} \right) \tag{3.5}$$

Plotting i_D as a function of V_{GS} and calculating the slope m give

$$m = K'_L \frac{W}{L} V_{DS}$$

The slope m can be estimated by choosing two points from the curve (Figure 3.46) and then the parameter K'_L can be deduced.

From the K'_L value, the low electric field mobility μ_0 can be calculated:

$$K'_L = \mu_0 C_{OX}$$

where C_{OX} has been determined in the first part.

3.4.6.3 *Determination of Body Effect Parameter* γ The influence of the bulk voltage on the threshold voltage is described by the body effect parameter γ (see Eq. 3.3). For each bias V_{BS}, the threshold voltage is different. As a first step, we will extract the threshold voltage V_T for different bulk–source voltages V_{BS}. As discussed in the previous section, Equation (3.4) represents a parabolic function. It has been shown that the extraction of the threshold voltage V_T can be done in straightforward manner by choosing two points, (i_{D1}, V_{GS1}) and (i_{D2}, V_{GS2}), on the characteristics $i_D(V_{GS})$. Furthermore, when the two points fulfill the condition that $i_{D2} = 4i_{D1}$, it follows that $V_T = 2V_{GS1} - V_{GS2}$.

The drain current i_D as a function of V_{GS} is represented in Figure 3.47 for different bulk–source voltages V_{BS} (−1, −2, −3, and −4 V).

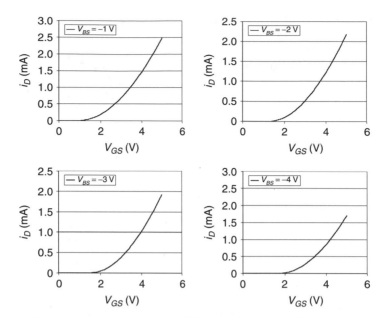

Figure 3.47 Drain current i_D as function of V_{GS} in saturation for different V_{BS}, $V_{DS} = 5$ V.

For each curve, the procedure described above can be applied and the threshold voltage extracted.

In the second step, the body effect parameter γ is determined. Equation 3.3 can be rewritten as

$$y = mx + b$$

which is the equation of a straight line. Identification of the parameters gives

$$y = V_T \qquad x = \sqrt{2|\Phi_F| + V_{SB}} - \sqrt{2|\Phi_F|}$$
$$m = \gamma \qquad b = V_{T0}$$

The parameter Φ_F can be calculated using the following expression and the already extracted parameter NSUB:

$$\Phi_F = \frac{kT}{q} \ln \frac{\text{NSUB}}{n_i}$$

Plotting $y = mx + b$ and calculating the slope give directly the body effect parameter γ (see Figure 3.48). The slope can be calculated by direct extraction by choosing two points from the curve.

To follow the described procedure, the student should first fill in Table 3.4.

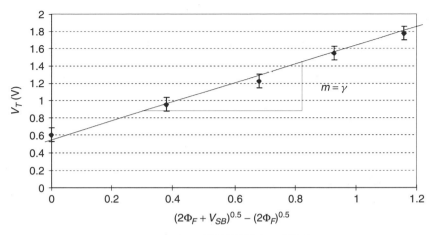

Figure 3.48 Plot of V_T as function of $\sqrt{2|\Phi_F| + V_{SB}} - \sqrt{2|\Phi_F|}$ to extract parameter γ.

Typical results are presented in Figure 3.48, in which the procedure has been applied to the measured results presented in this section.

3.4.6.4 Determination of Channel Length Modulation Parameter λ

The channel length modulation parameter λ describes the increase of the drain current in saturation with increasing V_{DS} and for constant V_{GS} and constant V_{BS}. This effect is also sometimes called the Early effect for MOS transistors, even when the physical origin for the Early effect in bipolar transistors is completely different.

To determine the parameter λ, Eq. 3.1 can be rewritten as

$$i_D = K_S' \frac{W}{2L} (V_{GS} - V_T)^2 (1 + \lambda V_{DS}) = i_{D0} + i_{D0}\lambda V_{DS} \qquad (3.6)$$

Plotting i_D as a function of V_{DS} and calculating the slope $i_{D0}\lambda$ and the y intercept i_{D0} from the data in the saturation region, λ can be determined by dividing the slope by the y-intercept value.

Figure 3.49 illustrates this procedure.

Table 3.4 Extraction of Body Effect Parameter γ

V_{BS} (V)	0	−1	−2	−3	−4
$y = V_T$ (V) (Use figures 3.45 and 3.47)					
$x = \sqrt{2|\Phi_F| + V_{SB}} - \sqrt{2|\Phi_F|}(\sqrt{V})$					

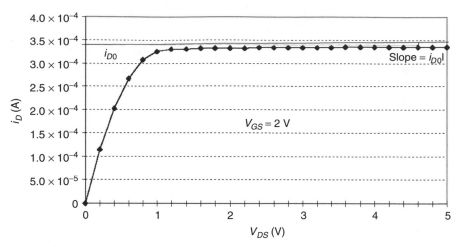

Figure 3.49   Plot of i_D versus V_{DS} to extract λ.

3.4.7   Measurement and Extraction

The characteristics $I(V)$ can be measured with the semiconductor parameter analyzer HP4155 as follows: This measurement instrument is located in Germany (University of Applied Sciences of Münster). On your personal computer, which must be connected to the Internet, start a browser program (Netscape or MS Internet Explorer). Load the URL http://www.retwine.net, click on Devices, and choose the instrument HP4155. In a first step, consult the tutorials to learn how to use the instrument.

When starting the measurement driver, wait until the front panel of the instrument is displayed on your PC screen (see Figure 3.22).

The device connected to the HP4155 in Münster is a MOS transistor (pins 1, 2, 3, and 5 of Figure 3.42).

The hardware connection is

SMU1 → Source, SMU2 → Drain, SMU3 → Gate, SMU4 → Substrate

Perform the measurements to extract the parameters V_{T0}, K'_L, K'_S, μ_0, γ, and λ.

3.4.8   Discussion

The transistors measured in this lab experiment have been realized within an "old" technology. Determine the characteristics of state-of-the-art technology in terms of

- Gate width
- Threshold voltage
- Oxide thickness
- Advantages to use such a technology
- Drawbacks.

3.5 EVALUATION

As noted earlier, the above lab exercises have been carried out by electrical engineering students at the University of Bordeaux 1, France. A questionnaire has been established to analyze students' reactions and to improve the RETWINE environment. The questions and answers are summarized in Table 3.5 (Geoffroy et al., 2001).

It seems that the virtual laboratory was accepted by the students, who realized that they were making real measurements. Furthermore, the virtual lab can be extended with additional lab exercises.

The complexity of the measurement instruments can be reduced by providing a tutorial or by developing user-scalable instrument interfaces. The real front panel could be replaced by a lab exercise specific interface, in which only the relevant stimuli must be entered.

The students seemed to indicate a need for improved communication; therefore, further communication tools should be developed and added to a virtual laboratory.

Finally, remote measurements break conventional frontiers and represent more than a new approach; they stand for a new dimension.

3.6 CONCLUSION

We have participated in an exciting new experience that consisted of making instrumentation available via the Web. The results can be summarized as follows:

- This project demonstrated the feasibility of web access for instruments.
- Some teachers have more problems using the new tool than the students.
- Internet connection problems occur sometimes, so users must be patient.
- The students were satisfied and enthusiastic, and they seemed to appreciate the new approach.
- The project made state-of-the-art, high-technology instruments available to students.

Table 3.5   Questionnaire for Virtual Lab Exercise

Q1:   Do you have the impression that you are taking a real practical course in electronics?

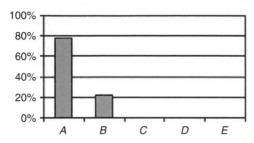

Q2:   Is the fact that the instruments are abroad motivating or without interest for you?

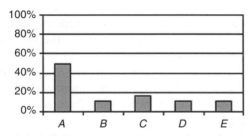

Q3:   Would you wish that other practical courses are put on-line?

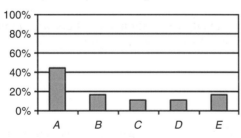

Q4:   Is the Internet interface for measurement user friendly?

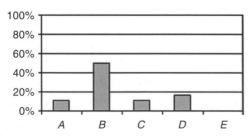

Table 3.5 *(Continued)*

Q5: The instruments are quite complex. Would you like to have a preliminary course on the use of each instrument?

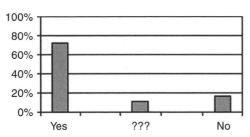

Q6: Did you prepare for the handling of the instruments using the on-line tutorial?

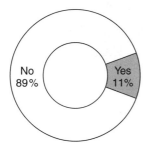

Q7: Did the fact of being in a PC room and not in a physical lab disturb you?

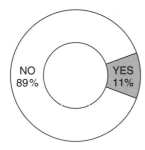

Q8: Would you like to make contact with the person in charge of instruments abroad?

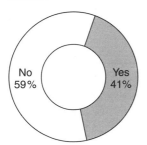

Table 3.5 *(Continued)*

Q9:   Which means of communication do you prefer: chat, email, telephone, fax, videoconference?

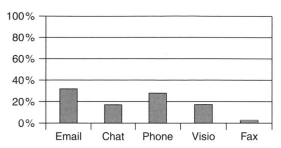

Q10: Do you feel ready to start a discussion in a foreign language? Which one?

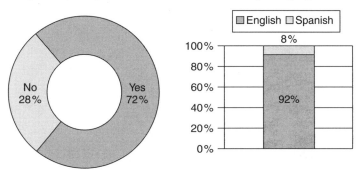

Q11: Do you think that making remote measurements adds a "European dimension" to the practical course?

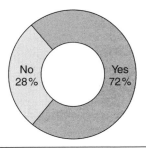

Question and response key: A = very well, B = well, C = no idea, D = unsatisfactorily, E = insufficiently

Extension of the instrumentation pool would result in a widespread international instrumentation lab, as other institutes provide additional instruments. In this way, we could create a complete virtual lab that makes the most modern instruments available to students.

ACKNOWLEDGMENTS

The RETWINE group would like to thank the European Commission for financial support via the Socrates—Open and Distance Learning Program.

REFERENCES

P. Allen and D. Holberg, *CMOS Analog Circuit Design*, Oxford University Press, New York, 1987.

M. Billaud, T. Zimmer, D. Geoffroy, Y. Danto, H. Effinger, W. Seifert, J. Martínez, and F. Gómez, "Real Measures, Virtual Instruments," in *Proceedings of the 2002 IEEE International Caracas Conference on Devices, Circuits and Systems*, April 2002.

M. Cervera, F. Gómez, and J. Martínez, "Remote Measurement System using Internet," in *Proceedings of the 1999 WebNet Conference*, Hawaii, October 1999.

D. Geoffroy, T. Zimmer, M. Billaud, Y. Danto, H. Effinger, W. Seifert, J. Martínez, and F. Gómez, "A Practical Course in a Virtual Lab," in *Proceedings of the 2001 EAEEIE, 12th EAEEIE Annual Conference on Innovations in Education for Electrical and Information Engineering (EIE)*, Nancy, France, May 2001.

F. Gómez, M. Cervera, and J. Martínez, "A World Wide Web Based Architecture for the Implementation of a Virtual Laboratory," in *Proceedings of the 2000 Euromicro Conference*, Vol. 2, Maastricht, Netherlands, September 2000, pp. 56–62.

Hewlett-Packard, HP 4145A Semiconductor Parameter Analyzer, *Operation and Service Manual*, Part No. 04145-90000, Tokyo, 1984.

Hewlett-Packard, HP 4284A Precision *LCR* Meter, *Operation Manual*, Part No. 04284-90000, Tokyo, 1988.

Hewlett-Packard, HP 8510B Network Analyzer, *Operating and Programming Manual*, Part No. 08510-90070, Santa Rosa, California, 1990.

Hewlett-Packard, HP 4194A Impedance/Gain-Phase Analyzer, *Operation Manual*, Part No. 04194-90001, Tokyo, 1991.

Hewlett-Packard, HP 4155A Semiconductor Parameter Analyzer, *User's Task Guide*, Part No. 04155-90010, Tokyo, 1994.

S. M. Sze, *Physics of Semiconductor Devices*, 2nd ed., John Wiley & Sons, New York, 1981.

Tektronix, GPIB-LAN Adapter, AD007. Available at `http://www.tek.com/Measurement/Products/catalog/ad007/60W_12024_1.pdf`, 1998.

T. Zimmer and D. Geoffroy, "Un exemple de TP à distance," *GeSi Revue des départements de Génie Electrique & Informatique Industrielle—IUT*, No. 55, pp. 18–22 (June 2000).

4

NEXT-GENERATION LABORATORY: SOLUTION FOR REMOTE CHARACTERIZATION OF ANALOG INTEGRATED CIRCUITS

C. Wulff, T. A. Sæthre, A. Skjelvan, and T. Ytterdal

Department of Physical Electronics, Norwegian University of Science and Technology, N-7491 Trondheim, Norway

4.1 INTRODUCTION

In this chapter we describe the development and use of the Next Generation Laboratory (NGL) (C. Wulff et al., 2002) that has been implemented at the Department of Physical Electronics, Norwegian University of Science and Technology, Trondheim, Norway. NGL is based on Microsoft's .NET technology and combines the latest in web technology with standard industrial instruments to make a cost-effective solution for education in the field of analog CMOS integrated circuits. NGL has evolved from work done elsewhere (Kristiansen, 1997; Dalager, 1998; Shen et al., 1999; Fjeldly et al., 1999, 2000; Shur et al., 1999; Smith et al., 2001). The main objectives for developing NGL are

Lab on the Web: Running Real Electronics Experiments via the Internet
Edited by Tor A. Fjeldly and Michael S. Shur
ISBN 0-471-41375-5   Copyright © 2003 John Wiley & Sons, Inc.

- To provide a remote laboratory course for education in the design of analog integrated circuits at our department.
- To create a platform for circuit experiments where it is easy to add new experiments.
- To create a prototype experiment that measures the frequency response of operational amplifiers that have been designed by students.

The outline of this chapter is as follows. In Section 4.2 we give an overview of the software technologies used. Then in Section 4.3 we describe the physical architecture and setup of the laboratory followed by a discussion of the software architecture employed. In Section 4.5 we describe the implemented prototype experiment, and in Section 4.6 we present a lab assignment given as part of the course Analog CMOS 1 that is offered to fourth-year students at our department. We end the chapter by discussing our experiences with NGL and future plans.

4.2 TECHNOLOGY

The development of NGL started in the summer 2001. At that time, several possible programming platforms for a remote laboratory were available, including active server pages (ASP), PHP: Hypertext Preprocessor (PHP), practical extraction and report language (PERL), .NET platform (beta 1), and several others. To select the programming platform that would suit us, we looked at the most critical portion of the application. The instruments we were going to use had, as do most commercial instruments, an implementation of the general-purpose interface bus (GPIB) through which one can remotely operate the instruments. This restricted our choices to a programming platform that could use the C++ library provided by the GPIB card. Although we could have used a number of programming platforms to interface with the GPIB C++ library, the .NET platform was emerging as a promising new technology and we wanted to explore its capabilities. The .NET platform has a large class library and seamless integration between standard Windows components and web applications. In addition, a new object-oriented programming language C# was released together with the .NET platform. The C# programming language combines a low learning threshold with the power of C++ and was therefore a prime candidate for the web application. The .NET platform has the advantage of being, in theory, both platform and language independent. When source code is compiled using a .NET compiler, the output is Microsoft Intermediate Language (MSIL). When a program runs, MSIL is compiled using a just-in-time (JIT) compiler, the same principle as Java. Since all languages produce MSIL, when compiled with a .NET compiler, MSIL entities can be used as components without the need to register them, as is the case of binary component object model (COM) components. Hence new features will emerge. An exciting consequence of this technology is the ability to write a class in C++ and inherit the class in C# or PERL. All these languages can make use of the extensive class library in the .NET platform. At the time of writing, several languages support the .NET platform, among them are VB.NET (Visual Basic 7.0), C#, C++, and PERL.

In addition to the server-side program necessary to perform the experiments, we needed a way to display the results to the user. The result from the prototype experiment consists of plots of the magnitude and phase of the output signal. To display these plots, we considered three solutions: creating a bitmap server side that is sent to the client, creating a Java applet to display the plots, or using a new technology from Adobe called SVG (scalable vector graphics). SVG was at the time implemented as a graph viewer at the LAB-on-WEB (Fjeldly et al., 2002) remote laboratory at UniK/NTNU, and we were fortunate to use this component and extend it to suit our purposes.

An explanation of the different technologies used in the NGL application follows.

4.2.1 .NET Platform

.NET Framework The .NET framework consists of the common language runtime (CLR), a JIT compiler, and a large class library. It is delivered in two versions, one with SDK (software development kit) and one without. Both can be downloaded free of charge from `http://www.asp.net/`.

ASP.NET ASP.NET enables rapid development of powerful web applications and services that can be viewed on any browser or device.

ASP.NET comes with a number of useful features for web developers. Among them, we have employed pagelets, web controls, and web services in this project.

Pagelets are reusable HTML code, which can be included into an ASP.NET page. They can be used as a container for common elements on HTML pages such as a menu or a header.

Web control makes it possible to create customized tags. Each tag has a class that defines its properties and what it displays on a web page. It is therefore an ideal solution for a system where several programmers are working on the same application. In the same way that a class can provide an easy-to-use interface for advanced logic, a web control can be used without knowing the intricate details of the inner logic.

Web service is application logic that is programmatically available over the Internet. Web services provide a programming abstraction that allows developers to easily make their applications available over the Internet without knowledge about HTTP, COM, TCP/IP, or data marshaling.

C# C# is a new object-oriented language developed by Microsoft. The reason for developing this new language was the lack of a sufficiently productive, yet flexible language. C/C++ has the flexibility, but developing new applications with C/C++ is too time consuming with the current demands on time to market. Microsoft's solution to this problem was C#, which possesses the productivity of Visual Basic without sacrificing the power and control of the C/C++ language.

The modern design of C# eliminates the most common C++ programming errors by applying

- Garbage collection that relieves the programmer from the burden of manual memory management
- Variables that are automatically initialized by the environment
- Type-safe variables

This makes it easier for developers to write and maintain applications.

4.2.2 XML

Extensible Markup Language (XML) is a markup language for documents containing structured information. The purpose of a markup language is to identify structures in a document. XML defines a standard, which does just that. In our application, XML is used for structuring the results from an experiment. Below is an example of the result information. Applying "tags" to the information elements makes the structures in XML. The properties of the experiment are the elements between the `<PROPERTIES>` `</PROPERTIES>` tags; the properties provide information on the parameters of the experiment. For example, the numbers between `<STARTFREQUENCY>` and `<STOPFREQUENCY>` are the start and stop frequencies of the frequency sweep. The second portion (`<GRAPH>`) contains all information used to draw the plots from the experiment (the X and Y values have been abbreviated to save space).

An XML Example

```
<?xml version='1.0'?>
<RESULT ID="Frequency Response AnCMOS" dateTime = "26.08.2002
13:03:21">
<PROPERTIES>
    <OPAMP>2</OPAMP>
    <RESISTOR>2</RESISTOR>
    <STARTFREQUENCY>10000</STARTFREQUENCY>
    <STOPFREQUENCY>40000000</STOPFREQUENCY>
    <BIASCURRENT>0.0001</BIASCURRENT>
    <VIPVOLTAGE>1.57</VIPVOLTAGE>
</PROPERTIES>
<GRAPH>
```

```
    <XVALUES>10000,10209.5159,... </XVALUES>
    <YVALUES>12.5311794782872,13.2404220730507,... </YVALUES>
    <XLABEL>Frequency</XLABEL>
    <YLABEL>Magnitude</YLABEL>
    <YVAL>dB</YVAL>
    <XVAL>Hz</XVAL>
    <VSIZE>250</VSIZE>
    <HSIZE>450</HSIZE>
    <TYPE_X>LOG</TYPE_X>
    <TYPE_Y>LIN</TYPE_Y>
    <SAMELINE>TRUE</SAMELINE>
</GRAPH>
<GRAPH>
    <SAMELINE>FALSE</SAMELINE>
    <XVALUES>10000,10209.5159,... </XVALUES>
    <YVALUES>-105.487563913513,-106.635818813394,... </YVALUES>
    <XLABEL>Frequency</XLABEL>
    <YLABEL>Phase</YLABEL>
    <YVAL>DEG</YVAL>
    <XVAL>Hz</XVAL>
    <VSIZE>250</VSIZE>
    <HSIZE>450</HSIZE>
    <TYPE_X>LOG</TYPE_X>
    <TYPE_Y>LIN</TYPE_Y>
</GRAPH>
</RESULT>
```

4.2.3 SVG

Scalable vector graphics (SVG) is a new graphics file format and web development language based on XML. SVG enables web developers to create dynam-
ically generated graphics on a web page. The main reasons for choosing SVG
are as follows:

- SVG is not a proprietary standard. This means that nobody owns it and
  everybody is free to use it in their products.
- SVG has zooming capabilities.
- SVG supports scripting, which enables a wide range of development possibilities.
- SVG is based on XML, and it is therefore possible to manipulate SVG
  files using standard class libraries.

SVG files tend to be very small. The main reasons are as follows:

- Shapes and characters are saved as text, not as graphics.
- Complex figures are defined by a <PATH> command. There is also a Bezier-curve command.
- You can define a symbol once and refer to it next time you need a similar symbol.

In SVG, the "painters" algorithm is used. This means that the current object is painted on top of the previous object. SVG can consist of three different types of graphic elements: shapes, text, and bitmaps.

4.2.4 CSS

Using cascading style sheets (CSS) is a simple way of defining in a centralized location how a page should look, thus avoiding the time-consuming job of setting, for example, font properties in every web page.

4.2.5 JavaScript

JavaScript was originally a platform- and browser-independent client-side scripting language. JavaScript was developed by Netscape but has been extended by Microsoft. This has led to a loss of browser independency, where some functions are not understood by some browsers. But as long as you limit yourself to the functions that are understood by all browsers, it is a good client-side scripting language.

4.3 PHYSICAL ARCHITECTURE AND EXPERIMENTAL SETUP

This section will concentrate on the physical architecture of the laboratory and the experimental setups. The physical architecture of NGL is shown in Figure 4.1. The architecture was based on the idea that NGL should be a scalable solution. The NGL web server runs the main web application that contains the application logic. On the next level, we have lab servers running the lab server web service. The lab server web service can be viewed as easy-to-use logic for accessing a GPIB and a data acquisition (DAQ) card installed on the lab server. Several instruments can be connected to the GPIB board. The instruments can be connected to a device under test (DUT). The instruments hooked up to a lab server can all be connected to the same DUT or they can be connected to different DUTs. This results in a great flexibility for specifying experimental setups.

The physical architecture of the NGL prototype, shown in Figure 4.2, is a subset of the architecture in Figure 4.1. It has one server that runs the web application and another server that runs the lab server web service. Connected to the lab server is a vector network analyzer from Rode & Schwarz, an Agilent

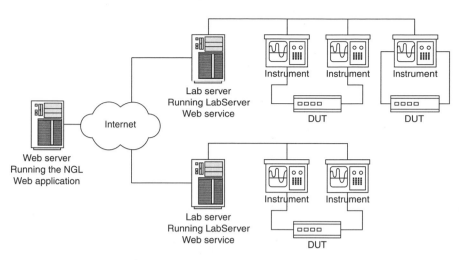

Figure 4.1 Physical architecture of NGL.

power supply, and a DAQ board. These are in turn connected to the DUT, the analog CMOS chip, which contains nine operational amplifiers (OPAMPs).

Fourth-year master's students at NTNU designed the OPAMPs in the fall of 2000 as a term project in the course Analog CMOS 1. The OPAMPs were integrated on the same chip together with some digital logic and an output buffer. The wiring diagram for the experimental setup is pictured in Figure 4.3. The prototype experiment allows the user to specify closed-loop gain, bias current to the OPAMP, and offset from the common mode on the positive input of the OPAMP. The resistors can be switched to give a closed-loop gain of 0, 20, 40, or 60 dB. The bias current can be varied from 0.5 to 25 µA, and the offset from the common mode can be varied from -100 to $+100$ mV.

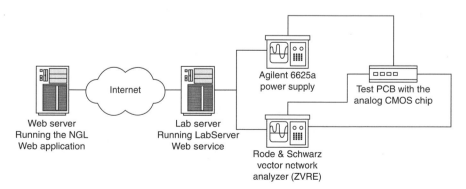

Figure 4.2 Physical architecture of NGL prototype.

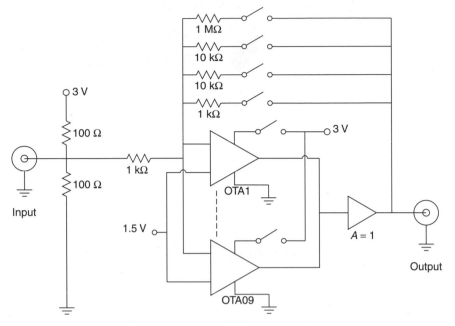

Figure 4.3 Analog CMOS chip test setup.

4.4 SOFTWARE ARCHITECTURE

The software architecture of NGL is divided into three parts (see Figure 4.4): web application, client-side graphics, and the lab server web service. The web application is divided into two components: presentation and logic. Presentation consists of the layout and the user interface of the NGL website. Logic contains the application logic that receives the parameters from the user, con-

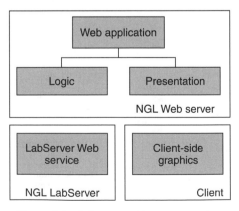

Figure 4.4 Software architecture overview.

trols the execution of the experiments, performs the experiment, and prepares the data for presentation. Client-side graphics contains the SVG control that is used to present the experimental results graphically on the client side. The lab server web service is, as mentioned, easy-to-use logic for accessing the GPIB and DAQ card for controlling instruments and the analog CMOS test board. Each part will be explained separately, starting with the web application followed by the client-side graphics module and the lab server web service.

4.4.1 Web Application Presentation

The layout and user interface of NGL are designed with simplicity and ease of use in mind. The front page is shown in Figure 4.5. It was very important that the layout should be easy to use and without "eye-candy" that would distract the user. The page is divided into three sections, the menu area on the left, the main area on the right, and a status field at the bottom of the page. The menu, CSS, and JavaScripts are included in the page using a pagelet. The experiments portion of the menu is automatically retrieved from the experiment classes in the application. We will show later how this is done. The status field provides the user with real-time feedback on the progress of the experiment. This feature

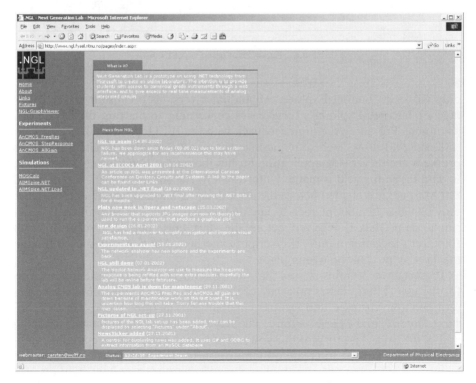

Figure 4.5 NGL front page.

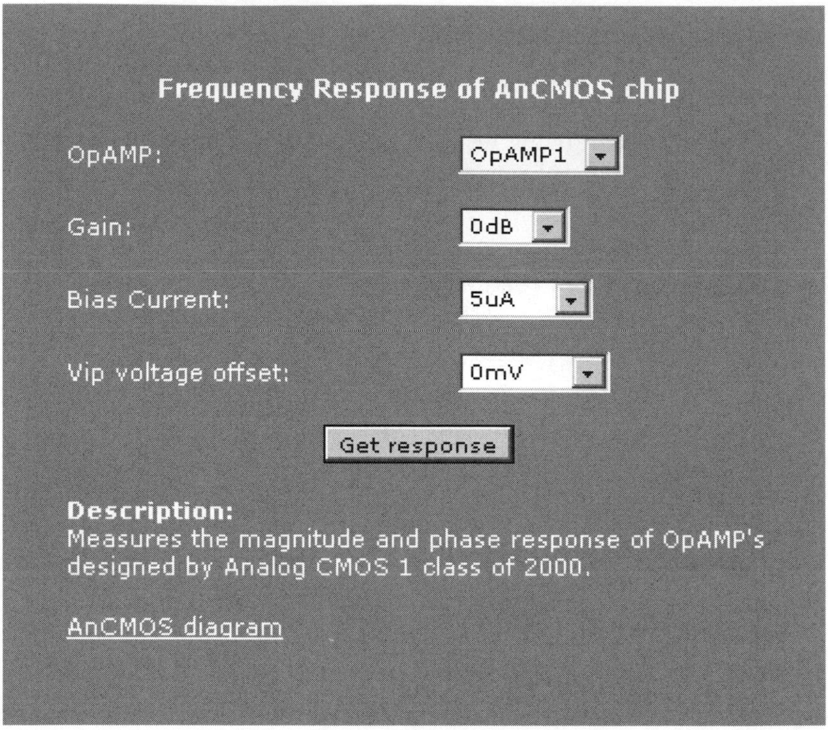

Figure 4.6   Experiment using `TextLayout`.

only works in Internet Explorer 5.0 and later versions because it takes advantage of Internet Explorer's forgiveness concerning formatting of the HTML document. Providing real-time feedback with the HTTP protocol is not an easy task, since it follows a client request–server response model. To circumvent this problem, the server-side application flushes the output buffer when a request to update the status field is made. Unfortunately, the start tags (e.g., `<HTML>` and `<HEAD>`) are not created before the experiment is finished, so the output buffer only contains JavaScript code to update the status field. Most browsers do not run a script that is received before the start tags, but because of Internet Explorer's forgiving nature, this script is interpreted and executed even though the JavaScript is in the wrong place.

Presentation contains classes that the experiment uses to create input forms for the user. These classes take care of the layout of the input forms, the values returned from the user, and browser capabilities. The three classes are `Layout`, `TextLayout`, and `GraphLayout`. The `Layout` class deals with event handling and is the base class for `TextLayout`. The `TextLayout` class contains methods that can be used to create an input form for the experiment, as shown in Figure 4.6. The `GraphLayout` class includes methods to allow creation of a more intuitive design of the form using background images and free positioning of form elements, as shown in Figure 4.7. Since it inherits `TextLayout`, stan-

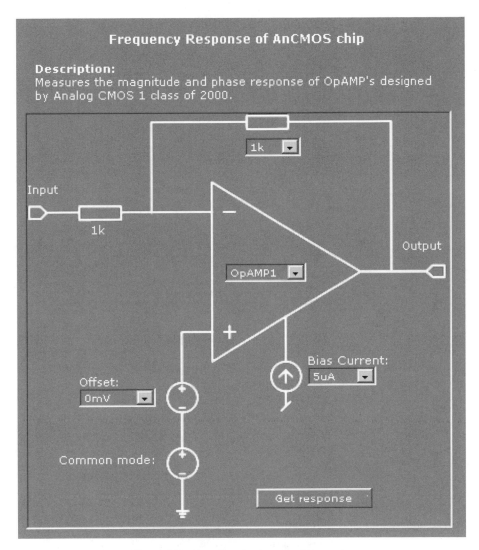

Figure 4.7 Experiment using `GraphLayout`.

dard form elements without free positioning can be requested. Figures 4.6 and 4.7 show the user interface of the prototype experiment, and we can easily see how Figure 4.7 provides a more intuitive interface for the user compared to Figure 4.6.

4.4.2   Web Application Logic

This subsection (and the following two) contains details on how the application logic was written. If you are not interested in programming, we would suggest skipping ahead to Section 4.5. However, if you intend to work on web laboratories or if you are a programmer, we believe that this subsection (and the following two) will provide you with some pointers.

Before we discuss the details of the web application logic, we will give an explanation on how web application works.

Figure 4.8 shows the flow of an experiment and how the different objects interact. The flow is somewhat simplified from the actual flow of the experi-

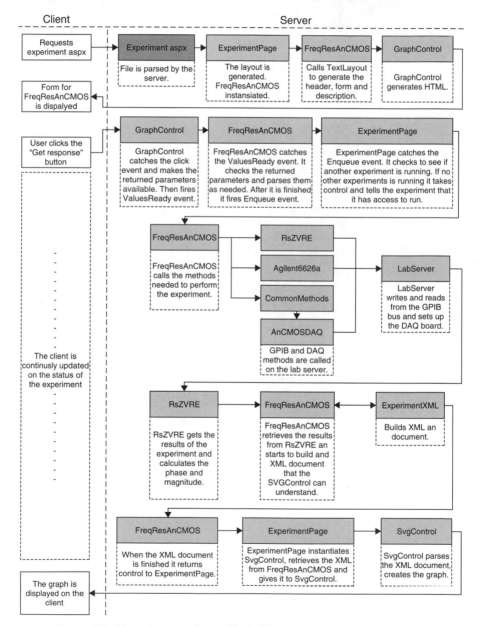

Figure 4.8 Flow of an experiment. Shaded boxes represent different classes.

ment but is sufficient for the discussion below. All the shaded boxes in the figure denote classes with the exception of `experiment.aspx`, which is a web page.

When the user selects an experiment, the page `experiment.aspx` is requested. Which experiment to draw is determined from the query string sent to the application. In this example we will use `FreqResAnCMOS`, which measures the frequency response of an OPAMP on the analog CMOS chip. The experiment utilizes the `GraphLayout` class to draw a form that contains the parameters the user can specify. `GraphLayout` makes use of a number of standard web controls in the .NET framework to ensure as much browser interoperability as possible. When the user is finished selecting the parameters, he or she clicks the `Submit` button. The click event is handled in the `Graph-Layout` control since the `Submit` button is one of its subcontrols. It parses the values returned and makes them available through a hash table. It then fires the `ValuesReady` event to notify the experiment that it has values ready for use. The `FreqResAnCMOS` object catches the event and the event handler calls `FreqResAnCMOS.setupExperiment()` to parse the returned parameters. When `ExperimentPage` has access to run an experiment, it calls the `Freq-ResAnCMOS.run()` method. The `FreqResAnCMOS` object utilizes the instrument objects shown in the figure but has no knowledge of where the actual instruments are located. After retrieving the raw data from the vector network analyzer, `RsZVRE` parses the data into a form that is easy for the `FreqRes-AnCMOS` object to use. The flow at this point is not event driven but follows the normal stack flow of the CLR. `ExperimentPage` retrieves an XML string that `FreqResAnCMOS` has built and gives it to `SVGControl`. Then, it returns control to the `experiment.aspx` page and the user sees the frequency response measured.

The back end of the web application is collected in a .NET assembly, called `NextGenLab`, which is a collection of classes much like a COM object. However, unlike a COM object, there is no need to register a .NET assembly. This makes it more manageable than COM and does not require a web server restart to be updated.

The class hierarchy of the `NextGenLab` assembly is shown in Figure 4.9. As we can see from this diagram, several of the classes inherit `WebControl`, all these classes can be instantiated directly in a web page by using a tag. The classes in `NextGenLab` are grouped in different namespaces, shown in Table 4.1.

NextGenLab   The classes in this namespace (see Table 4.1) are, with the exception of `BrowserCheck`, web controls that are instantiated by the NGL web pages. The `ExperimentPage` class handles what experiment to draw, controls the running of experiments, and displays the graph. When trying to implement a queue to prevent two or more experiments to run simultaneously, we experienced performance problems. With a queue in place, the overhead on performing an experiment was approximately 1–2 sec per experiment. This may not seem much, but the prototype was intended to be used as a lab for 30

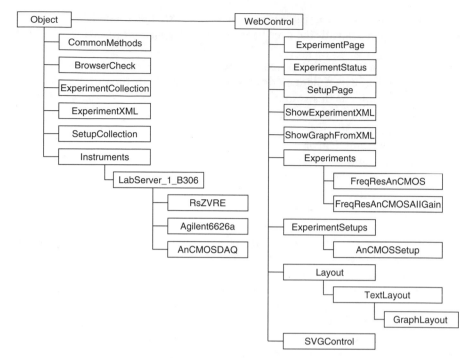

Figure 4.9 Class hierarchy.

Table 4.1 Namespaces and Classes in `NextGenLab` Assembly

Namespace	Classes
NextGenLab	BrowserCheck, ExperimentPage, ExperimentStatus, ShowExperimentXML, ShowGraphFromXML, SetupPage
NextGenLab.ExpControl	Layout, TextLayout, GraphLayout
NextGenLab.Experiments	Experiment, ExperimentCollection, ExperimentXML
NextGenLab.Experiments. AnCMOS	CommonMethods, FreqResAnCMOS, FreqResAnCMOSAllGain
NextGenLab.ExperimentSetups	ExperimentSetup, SetupCollection
NextGenLab.ExperimentSetups. Setups	AnCMOSSetup
NextGenLab.Instruments	Agilent6625A, AnCMOSDAQ, Instrument, LabServer_1_B306, RsZVRE

students, and you could easily end up waiting for a minute before the results are returned. The queue we implemented also had a tendency to hang if several experiments were run at the same time from the same computer. The queue idea was abandoned in favor of a fast and simple approach. In the NGL web application there is a global boolean variable that signals if an experiment is running. The `ExperimentPage` object checks this variable; if an experiment is running, it waits for 300 msec and tries again. When it finds that no experiment is running, it takes control and runs the experiment. To ensure thread safety, it locks the access to the global variable, sets the global variable, and unlocks the global variable. To give the user information on what the experiment is currently doing, it updates the status field provided by the `ExperimentStatus` web control. The classes `ShowExperimentXML` and `ShowGraphFromXML` are two web controls that allow the user to review the experiments performed in the session. In other words, you can return to results received 2 min ago and have another look. The stored results disappear when the user closes the browser window or the session time expires (20 min of inactivity).

The class `BrowserCheck` provides methods to check the browser capabilities and the browser type. `SetupPage` is a class that is essentially identical to `ExperimentPage` but controls the experiment setups.

NextGenLab.ExpControl   This assembly contains the classes explained under the web application presentation, which provides the experiments with methods to create a form with which the user can interact.

NextGenLab.Experiment   The `Experiment` namespace contains the framework for adding experiment classes. `Experiment` is the base class for all experiments, and all classes that inherit `Experiment` are automatically available through the menu. `Experiment` is an abstract class and cannot be instantiated directly. It contains four abstract methods that a child must implement. The abstract methods are `run`, `drawExperiment`, `Setup-Experiment`, and `IsOnline`. The `run` method should contain the code for performing the experiment, extracting the values from the instruments, and creating an XML string that is passed on to `SVGControl`, which is the web control that creates the plots. The `drawExperiment` method must contain the code for creating the form that the users utilize to interact with the experiment. In the spirit of simple creation of experiments, the `drawExperiment` method uses an instance of the `TextLayout` class to add textboxes, dropdown lists, and headers. An experiment can also choose to implement an overloaded version of the method `drawExperiment` that takes a `Graph-Layout` object as the input parameter, thus creating a graphical layout if the browser can handle it. The `SetupExperiment` method should parse the submitted values and prepare the experiment for running. The `IsOnline` method should implement the code for calling the `IsOnline` method in the instrument classes used by the experiment, which checks if the instruments are turned on and ready.

The class `ExperimentCollection` makes it possible to extract all classes that inherit `Experiment`. An example with this class is shown later.

The class `ExperimentXML` is a class that simplifies the creation of the XML string that `SVGControl` must have to draw a graph.

NextGenLab.Experiment.AnCMOS This namespace contains the classes for the experiments implemented in the NGL prototype. They are all derived from the `Experiment` class and contain the code for controlling the instruments and retrieving results from them.

NextGenLab.ExperimentSetup *and* ***NextGenLab.Experiment Setup.Setup*** This namespace contains the classes for providing a general setup for the experiments, that is, biasing the analog CMOS chip. This has been separated from the experiment to make the actual running of the experiment faster. If the same instrument is to be used in several different experiments, it is possible to implement the biasing and setup in the individual experiments.

NextGenLab.Instrument The `Instrument` namespace contains the classes for providing an easy interface to the instrument hardware. `Instrument` is the base class for all instruments and it mainly provides globalization of the decimal point such that a period is always used in conversion between string and double. `LabServer_1_B306` is a class that instantiates an object of the lab server web service and exposes the methods provided by the server that is connected to the actual instruments. The classes `Agilent6625A` (power supply), `AnCMOSDAQ` (DAQ board), and `RsZVRE` (vector network analyzer) provide easy-to-use methods for retrieving and setting values of the respective instruments. This limits the user's ability to perform illegal or damaging operations but also makes it easy for a user to interact with the instruments without knowing how to write GPIB strings. An example of retrieving the magnitude from the `RsZVRE` follows:

With the `RsZVRE` class:

```
RsZVRE rs = new RsZVRE();
rs.getResultsFromZVRE();
ArrayList xVals = rs.XValues;
ArrayList magn = rs.Magnitude;
```

Without the `RsZVRE` class:

```
HardInter hi = new HardInter(); //from proxy class
int device = hi.connectGPIB(20);
string xvals = hi.QueryGPIB(device,"TRAC:STIM? CH1DATA",7500);
string yvals = hi.QueryGPIB(device,"TRAC? CH1DATA",16000);
```

At a first glance, it might not seem too time consuming to get the magnitude without the RsZVRE class, but the *y* values returned by the second GPIB call are structured in a comma-separated string with alternating complex and real values. Therefore, to extract the magnitude, you have to separate the values and calculate the magnitude. This is, of course, not a difficult task, but having to write this code several times in different experiments makes it time consuming. The respective instrument classes also implement error handling.

ExperimentCollection and Adding Experiments As mentioned earlier, one of the main goals of the prototype was to make it simple to add experiments. Thus it was necessary to extract all experiments in the NextGenLab assembly and to choose at runtime what experiment should be run. Fortunately, the .NET framework makes this relatively simple. The class ExperimentCollection retrieves all classes that inherit Experiment through reflection when it is instantiated. The menu in the NGL application uses this class to retrieve names of the experiments in the NextGenLab assembly. Selecting an experiment is done by passing a GET variable to the ExperimentPage, that is, experiment.aspx?ExpID=1. The web control ExperimentPage uses ExperimentCollection.CreateInstance to create an instance of the experiment. This instance is casted to the Experiment class, but because of polymorphism, it retains its methods and variables. ExperimentPage treats this instance as an object of the Experiment class, which is the reason why an experiment must implement the methods mentioned earlier. If it does not implement these methods, the ExperimentPage object would not know how to interact with the experiment it just instantiated.

Here follows an example on the use of ExperimentCollection and polymorphism. The client requests the page experiment.aspx?ExpID=1:

```
public class ExperimentPage{
  .

  .
//Get index
int iExpID = Int32.Parse(this.Page.Request.QueryString("ExpID"));

//Create instance of ExperimentCollection
ExperimentCollection oExpCol = new ExperimentCollection();

//Get an instance of the selected experiment
Experiment oExp = oExpCol.CreateInstance(iExpID);
  .

  .
//Run The Experiment
oExp.run();
  .

  .
}
```

As shown in this example, the `ExperimentPage` object does not need to know what type of experiment it has instantiated, since they are all casted to `Experiment`. This works as long as all experiments have the same interface.

The use of reflection and casting in the NGL prototype has resulted in a very simple model for adding experiments. The procedure is shown in the following short recipe:

- Create a class that inherits `Experiment`.
- Implement four methods.
- Implement three static properties.
- Write the experiment logic in the `run()` method.
- Compile with the namespace `NextGenLab.Experiments.NameOfNewExperiment`.
- Copy the compiled dynamic link library (DLL) file to the bin folder of the NGL web application.
- Test the experiment through the NGL web application.

An experiment can be written in any of the languages that support the .NET platform.

4.4.3 Client-Side Graphics

The graph module was implemented as a web control; thus it is easy to reuse. The `SVGControl` web control parses an XML string and writes data to the browser needed to create the plot. A JavaScript is responsible for plotting the result by means of SVG.

SVGControl Class The class `SVGControl` is responsible for displaying the graphs. `SVGControl` receives an XML string from the experiment. The XML string is parsed and the results are sent to a JavaScript, which displays the results by means of SVG. The XML string contains information about the size, axis labels, and type (linear or logarithmic) of graph. `SVGControl` uses the size information to calculate the total height and width of the SVG object. An example of this XML file was given under 4.2.2 XML. Below you can see an excerpt from `SVGControl`. The `<embed>` tag is used to embed the SVG into the HTML page.

```
public class SVGControl : System.Web.UI.WebControls.WebControl
{
.

.

        protected override void Render(HtmlTextWriter output)
        {
        .

        .
```

```
            // Parsing the XML-string, writing values to
javascript.
            output.Write("<embed name='Graph' class=svgControl
            pluginspage='http://www.adobe.com/svg/viewer/
            install/' align='top' src='result.svg'
            width='"+max_width+"' height='"+max_height+"'
            type='image/svg-xml'>");


        .

        .

    }

.

.

}
```

The plotting and the interaction functionality of the graphs are obtained from JavaScript functions. Below is an overview of these functions:

Function **scale_and_plotaspx() {}**: This is the "main" function, which calls **plot()** for each graph.

Function **adjust_value_and_compute_prefix(axis_val_temp) {}**: This function adjusts the axis value and returns the correct prefix. SVGControl can handle values between 10^{-25} and 10^{27}. To improve readability, the axis values are adjusted. For example, 1,200,000 is adjusted to 1.2M.

Function **plot(k, r, gr, x0_outer, y0_outer) {}**: This function draws the frame, axes, labels, and gridlines. This function contains functionality to make the graph appear as consistent as possible. For example:

axis values after **adjust_value_and_compute_prefix()**: 0, 500m, 1, 1.5, 2, 2.5;

axis values after further adjusting in **plot()**: 0.0, 0.5, 1.0, 1.5, 2.0, 2.5.

As we can see, all values get the same prefix.

Function **prompt_x(evt) {}** and function **prompt_y(evt) {}**: These two functions are used to adjust the x and y axes manually. The function executes when the user clicks on the respective axis. The appearance of the prompt depends on whether the axis is linear or logarithmic. In the case of a linear axis, the user is prompted for the minimum value, maximum value, and step value. The step value represents the range between each gridline. In the case of a logarithmic axis, the user is only prompted for the minimum and the maximum decade. The user input is validated, and illegal values are rejected. To improve user friendliness, these functions understand both the comma and period as the decimal point.

Function **scale(min, max) {}**: This quite complicated function performs autoscaling. It returns new minimum and maximum values and an appropriate step value.

4.4.4 Lab Server Web Service

In the quest for finding a viable and robust solution to make the methods of the GPIB and DAQ card available to a .NET application, several solutions were explored. The initial idea was to use COM objects for the GPIB and DAQ board that exposed the methods available through the National Instruments C++ libraries. This turned out to be a solution that worked but was not nearly as robust as a web application demand. Using COM objects demanded that the NGL web application was run on the machine that had the DAQ and GPIB boards installed. Several attempts were made to make the COM solution more reliable and robust, but the attempts failed.

As the quest progressed, the use of a web service emerged as a possible solution. The DAQ and GPIB methods were implemented and exposed through a C++ web service. This turned out to be a very viable and robust solution. As a by-product of using a web service to interface with the GPIB and DAQ card, the ability to have several experimental setups connected to different lab servers spanning wide geographical areas emerged with no additional costs. In addition, the lab server is lightweight and can be run on an old computer. The NGL lab server is currently running on a 166-MHz Pentium processor with Windows 2000 Professional.

The prototype `LabServer` exposes five GPIB methods and five DAQ methods. Here follows the GPIB methods and an example:

GPIB Methods

```
int ConnectGPIB(int DeviceAddress);
int WriteGPIB(int DeviceHandle, string CommandStr);
int CmdGPIB(int DeviceHandle, string CommandStr);
string QueryGPIB(int DeviceHandle, string CommandStr, long
ReturnLength);
string ReadGPIB(int DeviceHandle, long ReturnLength);
int ReleaseGPIB(int DeviceHandle);
```

Use of Web Service GPIB Methods in C# In Visual Studio .NET adding a reference to a web service is as simple as adding a reference to a local COM or .NET component.

This example shows how to read the identity of a device connected to the GPIB bus at address 5:

```
using LabServer;
.
class LSTest{
.
public void Main(){
 Console.Write("Reading Identity: ");
```

```
//create an instance of the Web Service
HardInter oHInter = new HardInter();
string sCommandStr = "*IDN?"

//Connect to the device at address 5
int iDevice = oHInter.ConnectGPIB(5);

//Get the Identity
string sReturnedStr = oHInter.QueryGPIB(iDevice,sCommandStr,20);
oHInter.ReleaseGPIB(iDevice);
Console.WriteLine (sReturnedStr);
}
.
.
}
```

The console output if a Rode and Schwarz ZVRE was connected with the GPIB address 5 is as follows:

```
C:\>LSTest.exe
Reading Identity: Rode & Schwarz,ZVRE,8
C:\>
```

In theory any programming language can be used to access the web service since the method calls use HTTP and Simple Object Access Protocol (SOAP) in a normal request–response scenario.

4.5 FREQUENCY RESPONSE EXPERIMENT

As mentioned, the prototype experiment is devoted to measuring the frequency response of operational amplifiers. In this section, we will explain how to perform the experiment, show results from the frequency response experiment, and explain how to interpret the results.

To begin the experiment, we press the AnCMOS_FreqRes link on the left-side menu of the NGL opening page (see Figure 4.5). Figure 4.10 shows the start page for this experiment. Here we can specify the circuit parameters to use when running the experiment. For example, if we select 20 dB gain (10-kΩ resistor) and OPAMP 1, we will get the result shown in Figure 4.11. While the experiment is running, we get feedback from the server in the status line at the bottom of the browser. The status information only works with Internet Explorer 5.0 and later versions. In Figure 4.12, we show an example of how the status line might appear. The marked line indicates when the experiment was started (we pressed Get Response).

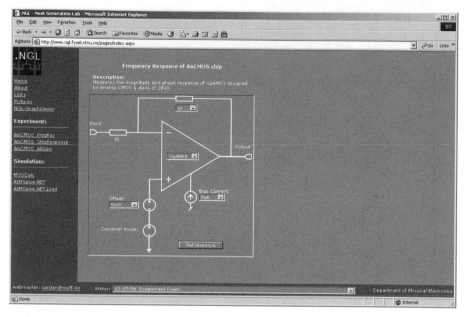

Figure 4.10   Start page frequency response experiment.

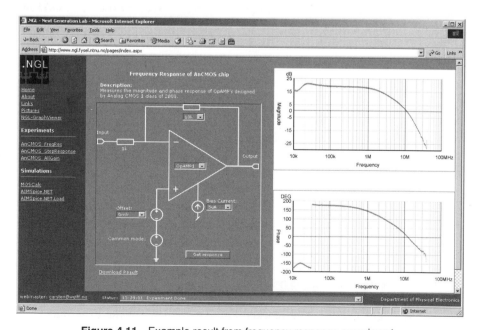

Figure 4.11   Example result from frequency response experiment.

Figure 4.12 Example of status line output.

From the results in Figure 4.11, we see that the frequency sweep starts at 10 kHz. This is because there is a DC stop filter connected between the test circuit and the vector network analyzer. The effects of this stop filter and the coaxial cables in the frequency range 10 kHz–40 MHz have been eliminated through calibration of the experimental setup.

We will now show how to determine the bandwidth (−3-dB frequency), phase margin, the 0-dB frequency (first-order equivalent to open-loop gain–bandwidth product), and some example experiments. We concentrate on OPAMPs 1 and 2 because these two are the most sensitive to change in offset and bias current.

To determine the bandwidth, we first find the −3-dB frequency. From Figure 4.13 we can see that the OPAMP has a stable gain at 19–20 dB, which gives a −3-dB frequency at around 2 MHz and hence a bandwidth of 2 MHz. The 0-dB frequency is at 6.5 MHz. To determine the phase margin in this case we find the phase at the 0-dB frequency (6.5 MHz). We can see that the phase margin is about 20 degrees. This is a very low phase margin and may lead to instabilities.

We choose OPAMP 2 to investigate what happens when we adjust the bias current to 0.5, 5, and 25 μA with 20 dB gain. We see from the magnitude plots in Figures 4.14, 4.15, and Figure 4.16 that the bandwidth and the 0-dB frequency increase as we increase the bias current. This is to be expected from the first-order equation of the unity-gain frequency:

$$w_t = \frac{g_m}{C_L} \qquad \text{where} \qquad g_m = \sqrt{2\mu_n} C_{ox} \frac{W}{L} I_d$$

The resonance peek in Figures 4.15 and 4.16 around the −3-dB frequency indicates instabilities of the OPAMP. To further illustrate the sensitivity of these OPAMPs and possible effects of an increase in bias current, we also study OPAMP 1 at bias currents of 5 and 25 μA with the gain set at 0 dB.

In Figure 4.17 we can see a resonance peak. The relatively low value of about 3 dB indicates that the amplifier still has some phase margin and is not very unstable. In contrast, the plot in Figure 4.18 shows a circuit configuration which is very unstable and has resonance peaks at 10, 20, and 30 MHz. These

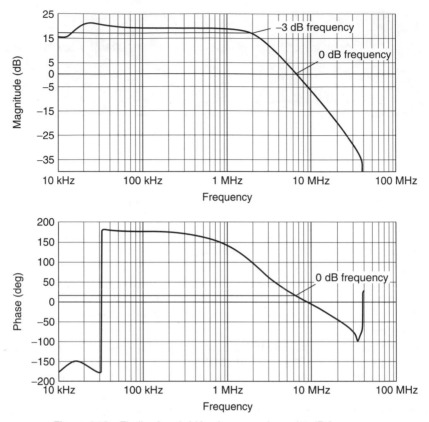

Figure 4.13 Finding bandwidth, phase margin, and 0-dB frequency.

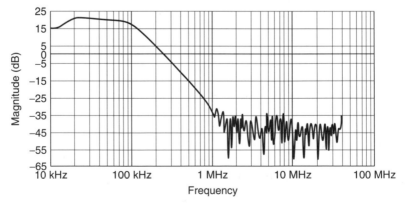

Figure 4.14 Magnitude response with bias current of 0.5 μA.

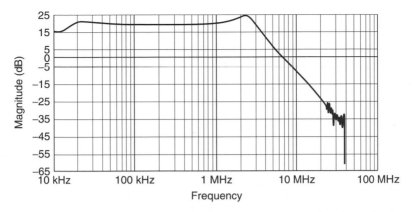

Figure 4.15   Magnitude response with bias current of 5 μA.

three peaks indicate that the circuit has a resonance frequency at 10 MHz while the first two harmonics are at 20 and 30 MHz. The gain of 40 dB is far beyond the open-loop gain at this frequency and thus another sign of oscillation. The high value is the result of a low phase margin, which results in positive feedback through the feedback resistor.

4.6 LABORATORY ASSIGNMENT

This section gives an example of a laboratory assignment given in the Analog CMOS 1 course at NTNU in the fall of 2001. To start the lab, go to `http://www.ngl.fysel.ntnu.no` and select `Experiment->AnCMOS_FreqRes`. *Note:* Almost all questions have short answers.

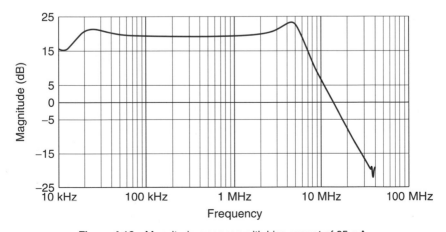

Figure 4.16   Magnitude response with bias current of 25 μA.

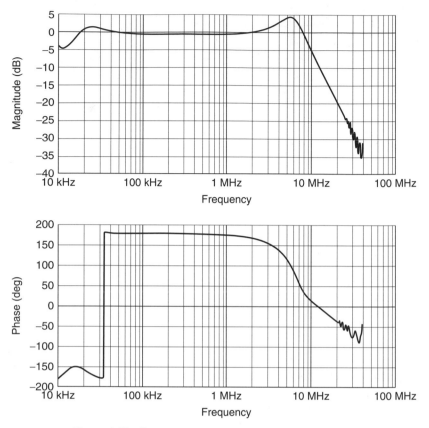

Figure 4.17 Frequency response with bias current of 5 μA.

Tips

- On the graph use Ctrl plus the left mouse button to zoom in, Ctrl plus shift plus the left mouse button to zoom out, or the right mouse button to show a menu.

Effects of Offset from Common Mode *Tip:* You can see the actual voltage value of the positive input displayed in the status field shown in Figure 4.19.

1a. Calculate the expression for the gain of the circuit in Figure 4.3.
1b. Select OPAMP 4 with 20 dB gain and run the experiment. You should see a proper frequency response. Change `Offset` to +100 mV and run again. What happened? Why?

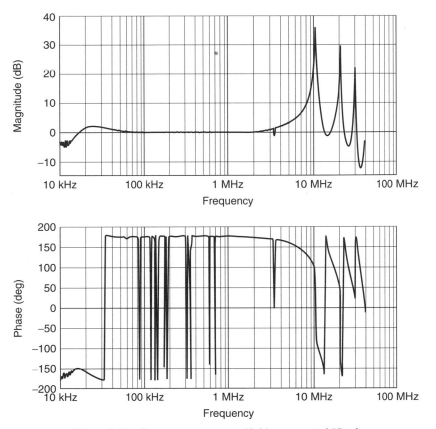

Figure 4.18   Frequency response with bias current of 25 μA.

1c. Change to +10 mV and run again. The frequency response should return. Change the gain to 60 dB and run again. What happened? Why?

1d. At what offset voltage should the AC gain "disappear" if you are using 20 dB gain, common mode 1.5 V, and a 3-V power supply? What about for 60 dB gain?

1e. Do your findings in 1d match what you saw in 1b and 1c?

Figure 4.19   Status field showing actual voltage applied.

Effects of Bias Current

2a. Select OPAMP 1 and gain 20 dB. You should see a similar frequency response as for OPAMP 4. Adjust the bias current up and down. What happens? Why?

2b. Select 0 dB gain, 5 μA bias current. Increase the bias current step by step. What happens? Why?

Phase Margin   Use estimates in these questions, for example, OPAMP 11 phase margin = 60–65 degrees.

3a. Estimate the phase margin of OPAMP 1 and OPAMP 2 using 20 dB gain.

3b. Is the phase margin in these two cases sufficient to ensure that the circuit never oscillates?

4.7   EXPERIENCES WITH NGL

At the time of writing, NGL has been running for 12 months. During this time, it has been used in the course Analog CMOS 1 at NTNU. The feedback from students has been positive, and it has underlined the effectiveness of a remote laboratory as a teaching tool. NGL provides students with an intuitive understanding of what happens when one adjusts parameters like bias current and offset from common mode on operational amplifiers. In addition, it underlines the importance of parameters like phase margin and it shows the effects when this parameter is too low.

The .NET platform has proven to be a very stable platform, in spite of the fact that from August 2001 to March 2002 the beta 2 version of the .NET platform was used.

4.8   CONCLUSION AND FUTURE PLANS

The NGL web application provides users with a reliable and efficient tool for analog integrated circuit experiments. It gives lecturers and students the opportunity to perform real-time experiments on actual circuits using industrial standard measurement equipment.

The NGL web application provides a framework for adding new experiments and experiment setups. Choosing the .NET platform as server-side technology provides distributed architecture with no additional cost. The NGL web application is scalable and easy to use.

The NGL is continuously updated and several plans exist for the future. At the time of writing, work is being done in three fields, time domain experiment, browser independency, and custom circuit setup with a programmable analog integrated circuit.

The time domain experiment measures the step response of the OPAMP circuit. The step response is used to measure the phase margin more accurately than the frequency response measurement.

To improve browser independency, a module is being developed that creates plots at the server side as bitmaps for browsers that do not support SVG. The server-side bitmap generator coexists with SVGControl, and the decision on which to use is done by checking the current browser's capability.

The long-term vision of NGL is to use programmable analog integrated circuits, thus providing a student with several circuits to choose from, for example, comparator, sample and hold, or DAC (digital-to-analog converter). These can be measured alone or combined into larger circuits like an ADC (analog-to-digital converter). A programmable analog integrated circuit named PAnIC (Wulff and Ytterdal, 2002) is scheduled for production in November 2002. A prototype of a custom circuit lab is scheduled for the summer/fall of 2003.

ACKNOWLEDGMENTS

The project was funded by the Nordunet Internet Technology in Laboratory Modules for Distance-Learning, principal investigator Tor A. Fjeldly, and the Department of Physical Electronics, NTNU.

REFERENCES

B. Dalager, "Remotely Operated Experiments on Electric Circuits over the Internet— Realizing a Client/Server Solution," M.Sc. Thesis, Norwegian University of Science and Technology, 1998.

T. A. Fjeldly, M. S. Shur, H. Shen, and T. Ytterdal, "Automated Internet Measurement Laboratory (AIM-Lab) for Engineering Education," in *Proceedings of 1999 Frontiers in Education Conference (FIE'99)*, San Juan, Puerto Rico, IEEE Catalog No. 99CH37011(C), Institute of Electrical and Electronics Engineers, Piscataway, NJ, 1999.

T. A. Fjeldly, M. S. Shur, H. Shen, and T. Ytterdal, "AIM-Lab: A System for -Remote Characterization of Electronic Devices and Circuits over the Internet," in *Proc. 3rd IEEE Int. Caracas Conf. on Devices, Circuits and Systems (ICCDCS-2000)*, Cancun, Mexico, IEEE Catalog No. 00TH8474C, Institute of Electrical and Electronics Engineers, Piscataway, NJ, 2000, pp. I43.1–I43.6.

T. A. Fjeldly, J. O. Strandman, and R. Berntzen, "LAB-on-WEB—A Comprehensive Electronic Device Laboratory on a Chip Accessible via Internet," *Proc. Int. Conf. on Engineering Education (ICEE 2002)*, Manchester, United Kingdom, pp. O337.1–5 (2002).

V. Kristiansen, "Remotely Operated Experiments on Electric Circuits over the Internet—An Implementation Using Java," M.Sc. Thesis, Norwegian University of Science and Technology, 1997.

H. Shen, Z. Xu, B. Dalager, V. Kristiansen, Ø. Strøm, M. S. Shur, T. A. Fjeldly, J. Lü, and T. Ytterdal, "Conducting Laboratory Experiments over the Internet," *IEEE Trans. Ed.*, Vol. 42, No. 3, pp. 180–185 (1999).

M. S. Shur, T. A. Fjeldly, and H. Shen, "AIM-Lab—A System for Conducting Semiconductor Device Characterization via the Internet," late news paper at 1999 International Conference on Microelectronic Test Structures (ICMTS1999), Gothenburg, Sweden, 1999.

K. Smith, J. O. Strandman, R. Berntzen, T. A. Fjeldly, and M. S. Shur, "Advanced Internet Technology in Laboratory Modules for Distance-Learning," in *Proc. American Society for Engineering Education Annual Conference (ASEE'01)*, Albuquerque, New Mexico, June 2001.

C. Wulff and T. Ytterdal, "Programmable Analog Integrated Circuit for Use in Remotely Operated Laboratories," in *Proc. International Conference on Engineering Education (ICEE2002)*, Manchester, August 2002.

C. Wulff, T. Ytterdal, T. A. Sæthre, A. Skjelvan, T. A. Fjeldly, and M. S. Shur, "Next Generation Lab—A Solution for Remote Characterization of Analog Integrated Circuits," in *Proc. Fourth IEEE International Caracas Conference on Devices, Circuits and Systems (ICCDCS-2002)*, Aruba, April 17–19, 2002, IEEE Catalog No. 02TH8611C, Institute of Electrical and Electronics Engineers, Piscataway, NJ, 2002, pp. I024.1–I022.4.

5

REMOTE LABORATORY FOR ELECTRICAL EXPERIMENTS

I. Gustavsson

Blekinge Institute of Technology, S-372 25 Ronneby, Sweden

5.1 INTRODUCTION

In this chapter the remote laboratory at Blekinge Institute of Technology (BTH) in Sweden is presented. The Internet provides the communication infrastructure between a student at home or elsewhere and the experimental setup in the laboratory. The three main objectives of the laboratory are

- To provide remote laboratory experiments to on-campus students as well as to distance-learning students as part of courses in electrical and electronic engineering
- To design remote laboratory exercises which are almost identical to local ones
- To use the equipment and the premises more efficiently than in a traditional laboratory

What is the main difference between traditional laboratory exercises and remote experiments in electrical and electronic engineering? You cannot perform any manual handling of components or test probes over the Internet or

Lab on the Web: Running Real Electronics Experiments via the Internet
Edited by Tor A. Fjeldly and Michael S. Shur
ISBN 0-471-41375-5   Copyright © 2003 John Wiley & Sons, Inc.

see what is happening with your naked eye. In a remote laboratory other types of devices should be considered than those normally found in university laboratories today. Computer-based instruments with virtual front panels and software drivers are appropriate for remote access. The drivers must support all the functions that those instruments provide because there are no other means to control them. They are not fitted with real front panels with control knobs.

At BTH the client–server paradigm is used, and in most cases several clients can conduct experiments on the same lab server simultaneously. A lab server is a PC with computer-based instruments and switch matrices controlling an experimental setup. The virtual front panels of the devices are included in the client software, and only settings and measurement data are transferred over the Internet. The basic TCP/IP are sufficient to transfer the small data streams involved. A 56-kbps modem in the client PC will do if video transmission is not required.

There are three lab servers in the remote laboratory at BTH. The first server used in the first-generation remote lab was used initially in a circuit theory course in September 2000. A remotely controllable switch matrix is used to form the circuits and to connect the test probes. The digital multimeter and the oscilloscope provided can detect all phenomena of interest. Neither video nor sound transmission is required. One of the experiments provided is a test of Kirchhoff's voltage law. Physical laws must be tested with real experiments. Simulations will not do. The client software needed and lab instructions can be downloaded from the laboratory home page.

The second server, the instrument server, is a more advanced version of the first and controls more instruments and switch matrices. It will later on replace the first one. The third, called the transducer experiment server, controls a mechanical transducer fixture. The transducers can be rotated and their performance tested in different directions. There is also a video transmission from a web camera looking at the fixture and associated equipment from the ceiling.

There are three second-generation client software packages. The first is used in a user's course on the function generator and the oscilloscope. These are basic instruments used in laboratory experiments in undergraduate education in electrical and electronic engineering. In the course you can learn how to use these instruments by practical training. The second package offers five electrical experiments. Three of them are the same as in the first-generation client. The third client package is more advanced and is used to conduct transducer laboratory experiments. These experiments allow both manipulations of the mechanical fixture via the transducer experiment server and control of electronic instruments and switches via the instrument server. The mechanical manipulation is a slow process that you can observe with your naked eye if you are present in the laboratory or through the web camera if you are conducting the transducer experiments from another location.

The second-generation client software packages are Visual Basic applications packaged for Internet deployment. The largest package is a little more than 5 MB. ActiveX technology is used to create components that can be embedded in an HTML page. This ActiveX solution can be downloaded with an Internet browser. To download a client software package from the laboratory website, you must lower your security settings temporarily since the software has no digital signature.

The second-generation system was used for the first time in an instrumentation course in December 2001. Only the second generation will be described in detail. Note that the servers are written in LabVIEW (Laboratory Virtual Instrument Engineering Workbench), and this software uses decimal points, not decimal commas, as in Sweden. For all inputs SI units are expected. The address of the remote laboratory at BTH is `http://www.its.bth.se/distancelab/english/`.

5.2 INSTRUMENTS FOR REMOTE ACCESS

In the local laboratories for undergraduate education in electrical engineering at BTH the instruments are ordinary desktop instruments with GPIB (general-purpose instrument bus, IEEE 488). Hewlett-Packard developed this interface bus at the end of the 1960s. The intention was to create a reliable bus system specially designed for connecting computers and instruments. GPIB is still the most widely used instrumentation connection interface, but its speed is sometimes not sufficient. Its maximum performance is 8 Mbps. Many desktop instruments have another disadvantage when you want to control them remotely from a PC running LabVIEW. In the drivers for many of these instruments only a subset of the functionality of the instrument is implemented. Thus for some settings you may have to use manual control buttons on the front panel of the instrument. However, new standards for instrument drivers are being defined within the IVI Foundation (Interchangeable Virtual Instruments), `http://www.ivifoundation.org`. The major instrument manufacturers formed this organization in August 1998.

The IVI architecture breaks the traditional instrument driver into two parts—an instrument-specific driver and a class driver. The instrument-specific driver functions the way traditional instrument drivers have in the past, but with underlying architecture that is optimized for performance and includes instrument simulation. The class instrument driver contains generic functions for controlling an instrument category and calls the corresponding instrument-specific driver functions directly. In our lab servers only instrument-specific drivers are used.

Computer-based instruments are well suited to remote labs. They consist of a hardware board to be plugged into a host computer and a software driver. An oscilloscope board from National Instruments is shown in Figure 5.1. There are no control buttons or display on its small front panel, only connectors. It has a virtual front panel that can be viewed on the computer screen. A virtual oscilloscope panel is shown in Figure 5.2. You have to use the mouse or the keyboard to do the settings. As there are no physical control knobs,

Figure 5.1   Oscilloscope board. (Reproduced courtesy of National Instruments Corporation.)

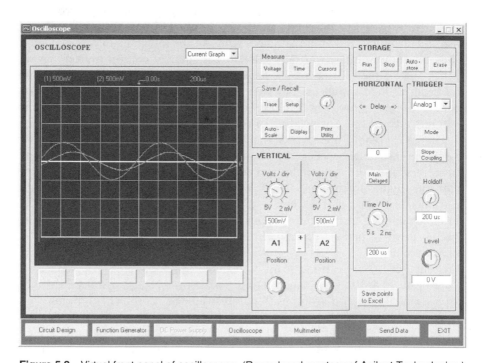

Figure 5.2   Virtual front panel of oscilloscope. (Reproduced courtesy of Agilent Technologies.)

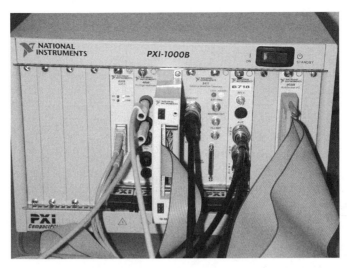

Figure 5.3 PXI box. Cables attached to leftmost socket are optical fiber cables connected to host PC.

software drivers for computer-based instruments must support all functions in the specification.

Plugging an instrument into a host computer, which is usually a PC, means that the instrument will be exposed to electric noise. A better way is to put it into a special low-noise expansion box designed for instruments. There are different standards for computer-based instruments. One of them is PXI (PCI Extensions for Instrumentation), `http://www.pxisa.org/`. National Instruments announced it in August 1997 as an open specification. It is built on the modular and scalable CompactPCI specification and the high-speed PCI bus architecture. A PXI box has slots for PXI instruments and for an embedded controller. The box has a PCI bus in the back plane. This PCI bus can be controlled by the embedded controller running Windows if the controller is present or connected to the PCI bus of a host PC through a special bridge unit. Then the instruments in the box will become devices under Windows. An example of such a box is shown in Figure 5.3. The number of different instruments designed to support the PXI standard and the number of suppliers on the market is still limited but growing.

5.3 WEB TECHNOLOGY

Internet Protocol (IP), User Datagram Protocol (UDP), and Transmission Control Protocol (TCP) are the basic tools for communication over the Internet. IP is used to perform the low-level service of moving data between computers. Data are packaged into components called datagrams. A datagram

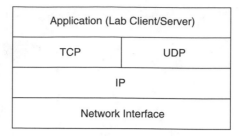

Figure 5.4   Protocol layering.

contains, among other things, data and a header that indicates the source and destination addresses. An IP software module determines the correct path for a datagram across the Internet and sends the data to a specified destination. IP cannot guarantee delivery. In fact, a single datagram may be delivered more than once if the datagram is duplicated in the transmission. Figure 5.4 shows the layered organization.

UDP provides means for communicating packets of data to one or more recipients but does not guarantee the safe arrival of the data to the destination. Furthermore, data sent in multiple packets may not arrive at the destination in the order they were sent. IP software handles the computer-to-computer delivery. Once a datagram reaches the destination computer, UDP software moves it to its destination port. If the destination port is not open, UDP software discards the datagram. UDP cannot be used in this application.

Software using TCP ensures reliable transmission across networks, delivering data in sequence without errors, loss, or duplication. TCP software will retransmit a datagram until it receives an acknowledgment. Five features characterize the interface between an application program and the TCP reliable delivery service:

1. *Stream Orientation.* When two application programs transfer data, we think of the data as a stream of bits divided into bytes. The data stream viewed as a stream of bytes is divided into segments for transmission.

2. *Virtual Circuit Connection.* Before the transfer can start, both the sending and receiving applications interact with their respective operating systems, informing them of the desire for a transfer. Conceptually, one application places a "call" which must be accepted by the other, much like a telephone call. Protocol software modules communicate by sending messages and verifying that the transfer is authorized and that both sides are ready. Once all details have been settled, the protocol modules inform the application programs that a connection has been established and that a transfer can begin. TCP permits multiple, simultaneous connections. A connection-based protocol is appropriate for clients accessing a remote lab server.

3. *Buffered Transfer.* To make transfer more efficient and to minimize network traffic, implementations usually collect enough data from a stream

to fill a reasonably large datagram before transmitting it across the Internet.

4. *Unstructured Stream.* The TCP stream service does not support structured data streams.

5. *Full Duplex Connection.* Connections provided by the TCP stream service allow concurrent transfer in both directions.

Every lab server in the remote laboratory initiates a connection (passive open) by waiting for an incoming connection. In establishing a TCP connection, the client software has to specify the address and a port at the server. A number between 0 and 65,535 represents a port. Different ports at a given address identify different services at that address. A lab server uses the TCP `Create Listener` function in LabVIEW to create a listener and uses the `Wait on Listener` function to listen for and accept new connections from clients. When a connection is established, the server uses TCP `Read` and TCP `Write` functions to read and write data to the other side. The client software packages use the Winsock control in Visual Basic.

A server and a client PC can terminate their conversation gracefully using a close operation. When an application program tells a TCP module that it has no more data to send, this module will close the connection in one direction only. To close its half of a connection, the sending TCP module finishes a segment, waits for the receiver to acknowledge it, and then sends a terminating segment. The receiving TCP module acknowledges the terminating segment and informs the application program on its end that no more data are available. Once a connection has been closed in a given direction, the TCP module refuses to accept more data for that direction. Meanwhile, data can continue to flow in the opposite direction until the sender closes it. Of course, acknowledgments continue to flow back to the sender even after a connection has been closed. In LabVIEW the TCP `Close Connection` function is used.

The second-generation client software is grouped into packages and stored on the laboratory website. These Internet packages contain all information needed for installation in a client PC. They are designed to be downloaded from a website. Internet Explorer uses a process known as Internet component download to install the client software on your computer. When a user accesses the web page that hosts the package, the system downloads it to the user's computer. The package is verified for safety, unpacked, registered, installed, and then activated. All of this occurs in the background and is controlled by the browser.

5.4 TRADITIONAL LABORATORY EXERCISES VERSUS REMOTE-ACCESS EXPERIMENTS

In the local laboratories for undergraduate education in electrical engineering at BTH there are eight identical lab stations equipped with desktop instruments and power sources. The same number of student teams can work in the laboratory supervised by one teacher. The teams use breadboards to form test circuits and to connect instruments and power sources. The cost of the equipment and its maintenance can be cut down if the number of lab stations is reduced and/or the laboratories are used outside ordinary working hours. However, most teachers wish to work during the day.

Computer-based instruments with virtual front panels can replace the desktop ones. Remotely controllable switch matrices can be used to form the circuits and to connect instruments and power sources. Then the whole experiment can be controlled from a PC and no manual handling of components or test probes is required. Performing traditional experiments, the students spend four hours in the laboratory and sometimes they use a significant part of that time to troubleshoot connecting failures due to a damaged breadboard. Nevertheless, students need a great deal of hands-on practice. The experience of sitting in the laboratory together with other students and with a teacher cannot be re-created by remote access. However, at BTH students gain hands-on practice in practical project courses. These provide the right methodology to cope with real-world problems such as EMC issues and finding bad connections. In such courses professional tools for circuit assembly are available.

In many experiments there are no sensations you can perceive with any of your five senses. It is not possible to observe the electric current with the naked eye or hear the electrons moving. There are, of course, exceptions, such as the incandescent lamp or lightning, where you can actually feel the heat and see the light caused by the electric current. To conduct such experiments remotely, neither video nor sound transmission is required. Instruments are normally used to observe what is happening and they produce only a small amount of data. Thus only a small amount of data has to be transferred over the Internet. A 56-kbps modem will do.

How do you know that these remote experiments are not a fake if you are sitting in front of a client PC at home? The answer is that you do not. It could be an advanced simulation. If you are standing close to the experimental setup in the laboratory but do not remove any connections or do other modifications affecting the performance, you will still not know if the experiment as it appears on a client PC screen nearby is a simulation or not. However, you may hear the clatter from the relays in the switch matrix unit when, for example, a remote client moves test probes. Maybe it would be nice to implement sound transmission only to let the remote user listen to the relays. Experienced people realize that such advanced simulation programs do not yet exist.

With appropriate commands a switch matrix can form a circuit and connect test probes much faster than a human being. It is possible for several clients to share one lab server if each client is allowed to make only one measurement at a time. For an oscilloscope this means one single shot only. A lab server with a switch matrix can do required connections in a fraction of a second, but the switch transients produced could be a problem. They must disappear before accurate measurements of desired quantities can be made. However, in electrical experiments in undergraduate education the time constants involved can be short—below 0.1 sec—without causing any inconvenience by selecting proper values for the components; the contribution to the response time will then not be significant. Thus on request from a client PC the server will first start to form the circuit and to connect the test probes. Then the server will do the settings of the sources and instruments. The relays will be engaged and the switch transients will appear and disappear. After a certain delay the circuit is expected to be in steady state. Then the oscilloscope will be armed and will wait for a trigger. The delay caused by the instruments will be low when computer-based instruments are used.

5.5   FIRST REMOTE CIRCUIT THEORY LABORATORY EXPERIMENTS AT BTH

The configuration of the first-generation setup for remote access at BTH is shown in Figure 5.5. A user can choose to conduct any one of five experiments in basic circuit theory. The first two experiments are tests of Kirchhoff's voltage law with DC and AC, respectively. The second, the AC circuit, is shown in Figure 5.6. Physical laws must be verified by real experiments. Simulations will not do. The last three experiments are filter measurements. There the students are expected to draw Bode plots. The virtual front panel of the oscilloscope is shown in Figure 5.2. Most of the functions of the 54600B oscilloscope from Agilent Technologies are implemented.

The student at the client PC selects an experiment and connects the test probes using the mouse. The test points available in Figure 5.6 are indicated in Table 5.1. Then the student makes the instrument settings using the virtual front panels offered by the client program, concluding by pressing a `Send` `Data` key in Figure 5.2 to send the data to the lab server. The lab server

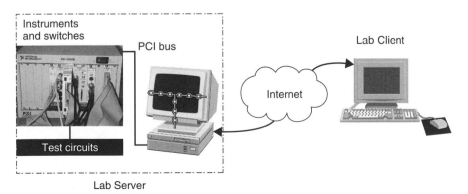

Figure 5.5   Configuration of first setup using computer-based instruments.

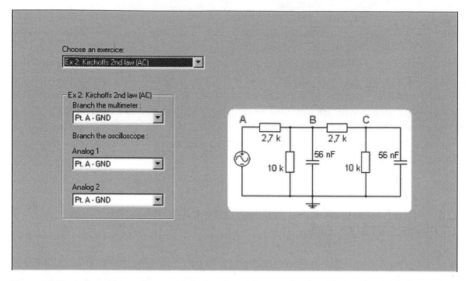

Figure 5.6 Laboratory options panel showing AC test circuits. There are three test points available: A, B, and C. Analog 1 and analog 2 are the oscilloscope input channels 1 and 2, respectively.

forms the required circuit and connects the test probes using a switch matrix. Then the lab server makes the settings requested and reads the instruments. Finally, the lab server returns the results obtained.

In the diagram in Figure 5.7, the switch matrix, the instruments, and a prototype board containing the test circuits are shown. The digital-to-analog (D/A) converter in the multifunction input–output (I/O) board is used as a DC power supply to feed the test circuits. You can find the test circuit of Figure 5.6 in the diagram. It is possible to do floating measurements with the digital multimeter. Such measurements could not be made using ordinary desktop oscil-

Table 5.1   Test Points in Tests of Kirchhoff's Voltage Law

Instrument	Test Points in Figure 5.6
Multimeter	A–Gnd
	B–Gnd
	C–Gnd
	A–B
	B–C
Oscilloscope	A–Gnd
	B–Gnd
	C–Gnd

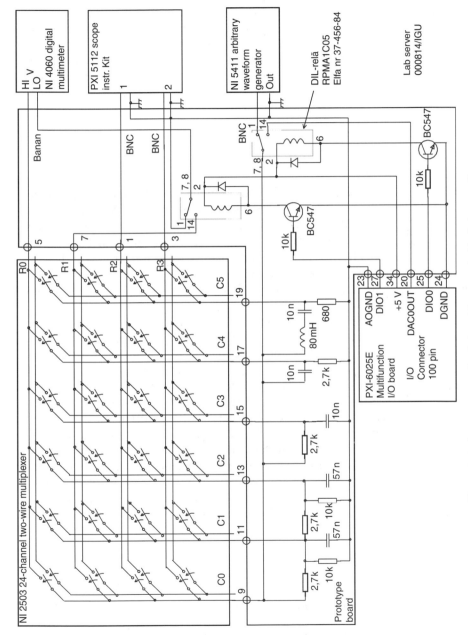

Figure 5.7 Instruments and switch matrix connected to first-generation server.

loscopes or a PXI oscilloscope. In traditional laboratory exercises some students try to do floating measurements with the oscilloscope, and it should perhaps be possible for them to try it also here. Compared to a circuit on a breadboard, a circuit formed using a switch matrix requires in most cases additional cabling because the switch matrix is on a board inside the PXI box and the components forming the circuit on another board are somewhere outside. This may limit the frequency range. To avoid extra cables and increase the frequency range, the switches should be located on the same board as the components.

5.6 INSTRUMENT SERVER

The instrument server controls a digital multimeter NI 4060, a function generator NI 5411, a function generator NI 5401, an oscilloscope NI 5112, a multifunction I/O board PXI-6025E, and two switch matrices NI 2503 in a PXI box connected to the server PC via an interface module MXI-3. The software is written in LabVIEW version 6i. The drivers used are NI-DMM 1.6, NI-FGEN 1.5, NI-SCOPE 2.0, DAQ 6.9.1, and NI-switch 1.6. Hardware and software manuals can be downloaded from the National Instruments website (www.ni.com/manuals).

To cope with requests from several users simultaneously, the server must have a short response time but also a queue manager to take care of simultaneous calls. In the LabVIEW full development system there is an example Date Server.vi in the file tcpex.llb which includes a queue manager. In Figure 5.8 a listing of the top layer of the server program is shown. It is simplified and the error processing is omitted. At startup a listener is created to listen for a client trying to open a connection. If the port number is correct, every connection attempt is accepted. A connected client is allowed to enter a "waiting room". If no one is in the laboratory, the client will be allowed to pass into the laboratory and perform a single measurement and leave. If someone is in the laboratory, the client has to wait in line if there are other clients waiting. Every time a client is allowed to pass into the laboratory, a message is sent to each client in the waiting room. When connected a client is expected to send two types of messages to the server:

- *Information Request.* The server response will be a list of the features of the server.
- *Setup and Execute Request.* The message is used to control the instruments and the switches plugged into the server. Since several clients can use the server simultaneously, all instruments and switches to be used must be set up at each call. The message is divided into packets—one packet for each instrument or switch.

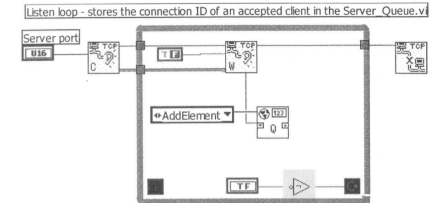

Listen loop - stores the connection ID of an accepted client in the Server_Queue.vi

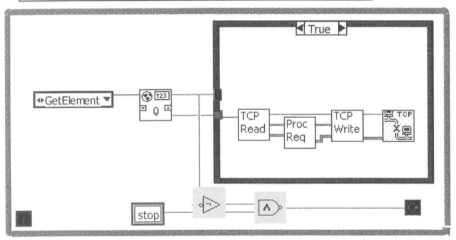

Work loop - let the first user in the waiting room conduct one experiment

AddElement - let the requesting client enter the waiting room

GetElement - let the first waiting client enter the lab and conduct one experiment

Figure 5.8   Top layer of instrument server.

The server will return one of three types of messages:

- *Queue Message.* The server returns this message if it is busy. The message contains only the queue position.
- *Error Message.* If an error occurs, the server returns an error message.
- *Data Message.* This message is divided into packets containing data output from some of the instrument used. The instruments producing output are the oscilloscope and the digital multimeter. The packets are in the same order as in the corresponding request message.

The packets in the `Setup` and `Execute` message contain the settings for the instruments and switches. For the digital multimeter, for example, the packet contains five settings supported in the server. Those are `Function`, `Resolution`, `Range`, `Number of Measurements`, and `Autozero`. The Lab-VIEW virtual instrument handling the digital multimeter is shown in Figures 5.9 and 5.10. The packet is decoded in the server. LabVIEW uses the C language notation `%d`, `%f`, and so on, to decode the strings.

The oscilloscope is somewhat more complicated. There are more settings and it produces more data. Most functions provided by the NI 5412 board are implemented. The server supports only single shots, but a client may repeat its request and get more than one trace pair per second if the server is not loaded. The NI-SCOPE driver provides read functions and fetch functions. The read functions are the easy way to acquire data, but the fetch functions offer an advantage. Fetching refers to the process of transferring the acquired waveform from the NI 5112 board memory to the host computer memory. The advantage is that the fetch functions allow you to fetch binary data instead of slower-to-acquire scaled data. The transfer is generally done with direct memory access, which copies the binary data from the board memory quickly. The vertical resolution of the NI 5412 board is eight bits and the fetch function selected produces an array of unsigned integers (bytes), and they are converted to a string to be sent to the client program. Each array element could be interpreted as an ASCII value. Assume that the height of one division on the oscilloscope screen is 1 cm. Vertically there are eight divisions. The vertical resolution is $80/250 = 0.32$ mm or approximately only three dots per millimeter. This should be enough for trace to be viewed on a screen. In the horizontal direction there are 10 divisions, and to get the same resolution on the screen in that direction 300 samples are required. Currently the server is producing 500 samples per trace. The board memory is much longer. Thus, if both traces are enabled, the oscilloscope produces only a little more than 1 kbyte per access.

There are two function generators and one of them supports the burst mode. Both generators can generate arbitrary waveforms. The waveform memory is 16,384 points. The server uses only 500 points for user-defined waveforms. The waveform data must be scaled between -1.0 and 1.0 in floating-point format. The amplitude parameter is used to set the actual output voltage.

5.7 TRANSDUCER EXPERIMENT SERVER

The transducer experiment server controls, among other things, the rotation of a mechanical transducer fixture shown in Figure 5.11. It takes some time to rotate the fixture and only one client at a time can use this server. There is a login procedure to give you exclusive right to control it. Currently you can use the server for 30 min, but you will be logged out if you do not use it for 10 min. The server also controls three fluorescent tubes on tables in the laboratory and the ceiling lamp. The experimental setup is shown in Figure 5.12.

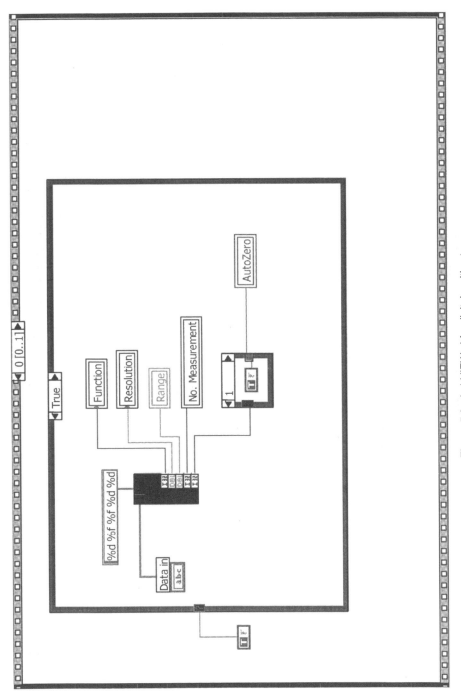

Figure 5.9 LabVIEW vi for digital multimeter.

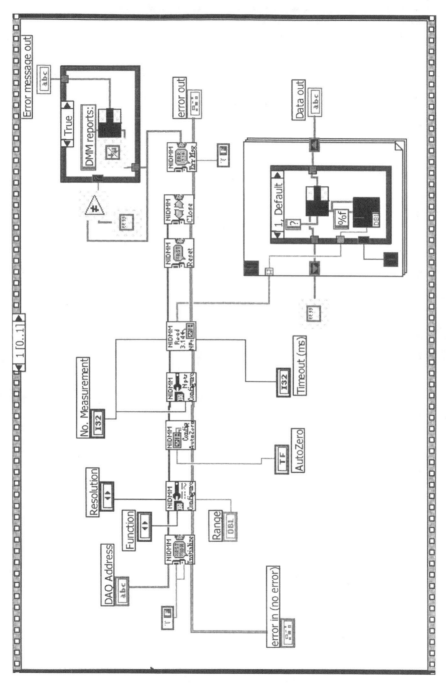

Figure 5.10 LabVIEW vi for digital multimeter.

Figure 5.11 Transducer fixture. Camera and two ultrasound transducers are attached to stepper motor shaft.

A client is expected to send three types of messages to the server:

- *Client password.*
- *Command.* Only three bytes containing the angle to rotate the fixture and on/off status for the lamps.
- *Kill request.* In return to a server message the client program sends the kill request to tell the server that the transmission is completed and to ask the server to close the connection.

The server will return one of seven messages:

- *Access Granted.* If the password is valid, then the server will return the number of minutes you are permitted to use the server and the number of minutes you are allowed to be inactive without being disconnected. The client program can then start sending commands to the server.
- *Wrong Password.* Access is denied.

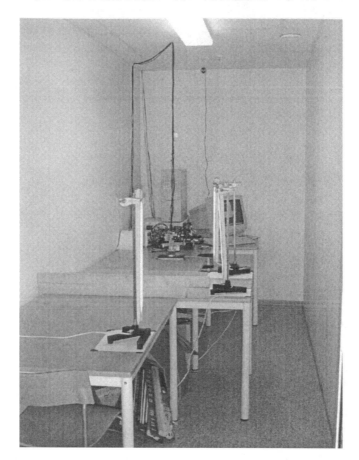

Figure 5.12   Laboratory is a small room without a window.

- *Server Busy.* The server will return the number of minutes left for the user logged in and the IP address of that client PC. The new client PC will get waiting status and will be logged in when the server becomes free if still waiting.
- *Client Waiting.* The server will return this message if one client is using the server and one is waiting. Access is denied.
- *Time Out.* Time is up or inactivity time out.
- *Error Message.* If an error occurs, the server returns an error message.
- *Ready Message.* The server is ready for a new command.

5.8 INSTRUMENT USER COURSE CLIENT PACKAGE

This client software package is available for download for students and other people who want to get experience of how to use two of the basic instruments

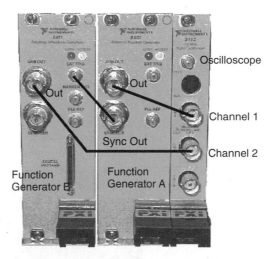

Figure 5.13 Instrument connections in user's course.

in the laboratory—the function generator and the oscilloscope. The student is expected to be familiar with basic electricity and concepts like voltage, current, impedance, phase shift, and sine- and square-wave signals. After the course the student is expected to be able to use such instruments.

Two function generators and an oscilloscope are connected as is shown in Figure 5.13. The Sync Out signal of generator A is connected to the Ext Trig input of generator B. The signal produced by generator A can then, for example, be used to start a burst in generator B or the phase shift between the two output signals can be controlled.

The virtual front panels of the three instruments are shown in Figure 5.14. The bottom half is the oscilloscope. The student can use the panels to do all settings, but a message containing the settings will be sent to the server only when one of the keys Start or Single in the top right corner of the oscilloscope panel is pressed. If you press the Single button, you will get data from a single shot only, but if you press the Start button, the client will, during a limited time, continuously send the message to the server. The caption of the Start key will then be changed to Stop, and the oscilloscope trace will almost immediately reflect settings changes done and you will get the same impression as if you were using ordinary desktop instruments in a laboratory. You are only allowed to use this continuous mode during a short period of time. Otherwise the response time will increase for other connected clients.

Beneath several of the buttons is a white text box. It shows your desired settings to be sent to the server. If, for example, you change the scale in the y direction with the Volts/Div knob and then press the Single button, the only thing changing immediately is the value in the text box beneath the knob. The displayed trace will not change until the server receives your setting and

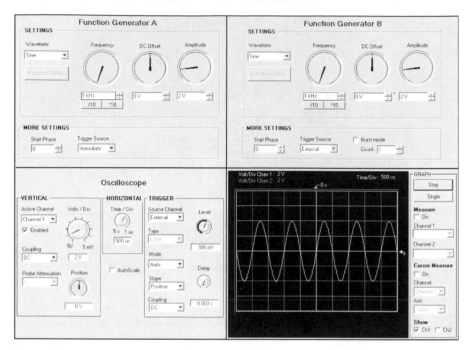

Figure 5.14   Course window showing virtual front panels of oscilloscope and two function generators.

returns a new trace in the new scale. The scale of the display is shown in green text in the top left corner of the oscilloscope display.

Some students claim that the output of the function generators is twice the value they expect. These students do not know that function generators with a normal output impedance of 50 Ω are designed to be loaded with 50 Ω. Unloaded they will output twice the voltage set. The input impedance of the oscilloscope is high so you can consider the function generators as unloaded.

5.9   CLIENT PACKAGE FOR ELECTRICAL EXPERIMENTS

The electrical experiments are similar to those in the first-generation circuit theory laboratory. One of the experiments shown in Figure 5.15 is a bit special. If you try to break a current through a solenoid, a high voltage will be generated. This phenomenon is used in the ignition system of a car. If you want to simulate it, you might get into trouble because, theoretically, the voltage generated across the solenoid will be infinite. However, if you do the real experiment in Figure 5.15 and open the switch 0.5 sec after the circuit is formed, the voltage in test point B will not be infinite. In fact, this is a second-order system, and the current will go through the oscilloscope. If you take the oscilloscope

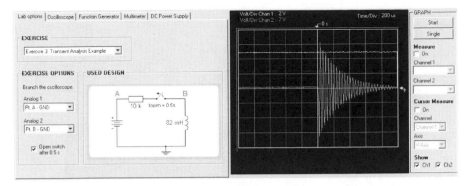

Figure 5.15   Exercise 3-a current through solenoid is broken.

impedance into account in the simulation, you will get a better result. However, the cables should also be considered.

5.10   CLIENT PACKAGE FOR TRANSDUCER EXPERIMENTS WITH VIDEO TRANSMISSION

BTH offers a project course where teams of students are expected to design and implement control systems for small vehicles. At the end of the course there is a contest where the vehicles must navigate autonomously through a racetrack filled with obstacles in the direction of a guiding light. The vehicle is fitted with transducers, a line scan camera, and two ultrasound transducers. It should find its bearing to the light source with the aid of the line scan camera and detect the obstacles in its way using the ultrasound transducers. The exercise requires that students produce some electronics and software. In the camera box there is only one linear sensor chip; the sensor array is mounted horizontally in the focal plane.

The students are invited to conduct remote experiments to determine how the transducers can be used to help the vehicle find its way to the light source and avoid the obstacles. In the experimental setup the camera box and the ultrasound transducers are attached to a stepper motor shaft, as shown in Figure 5.11. The experiments are arranged in a closed room without windows but with a network connector and a mains outlet, as is shown in Figure 5.12. The transducers and the stepper motor are fed through cables from the ceiling. In the room there are three light sources in addition to the ceiling lamp and one obstacle, a plank. Two of the walls can also be used as obstacles.

The transducer experiment server controls the transducer fixture and the light sources. The students are only allowed to rotate the fixture within a 360° range. There is a rotation limit in the server software. The instrument server controls the two function generators, the oscilloscope, and a relay switch

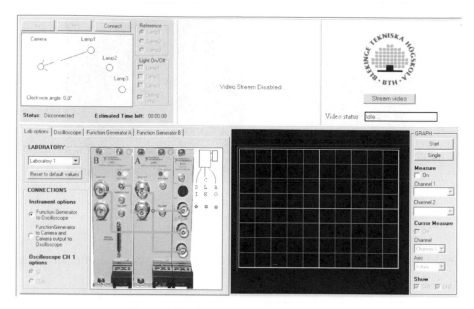

Figure 5.16   Transducer experiments window.

matrix used to make the required connections. A third server controls the web camera looking at the scene from the ceiling.

The client program window at startup is shown in Figure 5.16. The bottom part of the window contains the virtual front panels of the instruments and the lab options panel controlling the switch matrix. A drawing of the matrix is shown in Figure 5.17. When the client is started, the lab options panel and the oscilloscope screen are displayed. There are two lab assignments, line scan camera experiments and one where you are expected to design a range finder. The top left part of the window is used to control the transducer experiment server.

5.11 LABORATORY EXERCISES

5.11.1 Electrical Experiments

There are five experiments:

1. *Exercise 1.* Test Kichhoff's voltage law (KVL) with DC. According to this law, the potential difference around a closed circuit is always zero.
2. *Exercise 2.* Test KVL with AC.
3. *Exercise 3.* Show what happens when the electrical current through an inductor is broken.

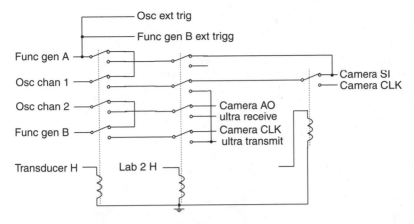

Figure 5.17 Switch matrix for transducer exercises.

4. *Exercise 4.* Determine what is in the black box.
5. *Exercise 5.* Examine a serial resonance circuit.

Preparatory Questions for Exercises 2 and 5

1. Draw a phasor diagram for U, U_1, U_2, U_3, and U_4 in Figure 5.18; $|U| = 8$ V_{pp} and phase angle 0.
2. Calculate the resonant frequency f_0 and use, for example, PSPICE to draw a Bode plot for the circuit.

Experiments

1. Use the circuit in Exercise 1 to test KVL. The circuit is shown in Figure 5.18. Set the DC power supply output to 4 V. Measure the voltages U_1,

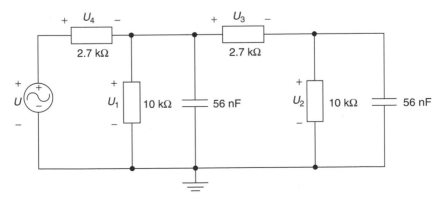

Figure 5.18 Circuit for test of Kirchhoff's voltage law.

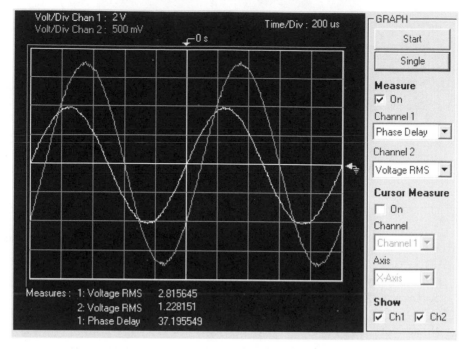

Figure 5.19   Oscilloscope settings for phasor measurements.

U_2, and U_3 with the multimeter. Calculate $U_1 - U_2 - U_3$. The capacitors will behave as an open circuit with DC.

2. Go to Exercise 2. You are now expected to test the same circuit with AC. Set the function generator to output a sine wave of 1 kHz and 8 V_{pp}. Repeat the same measurements and calculation as in Exercise 1. The result is not zero, but according to KVL, it should be zero. The reason for this is that the multimeter does not consider the relative phase between the various voltages but instead only shows the effective AC value. Measure instead the phasors U, U_1, and U_2 and use the phase of U as a reference phase. Figure 5.19 shows how you can use the oscilloscope to measure the complex values. Draw a phasor diagram for U, U_1, and U_2. Measure the length of the phasors U_3 and U_4 in the diagram. Also measure $|U_3|$ and $|U_4|$ with the multimeter and compare with the length of the corresponding phasor. Normally you will get differences because of the measurement uncertainties. To get the best accuracy from the phase shift computing algorithm and the analog to digital (A/D) converter in the oscilloscope, try to show only a little more than one period on the oscilloscope and the traces should almost hit the bottom and top of the oscilloscope display.

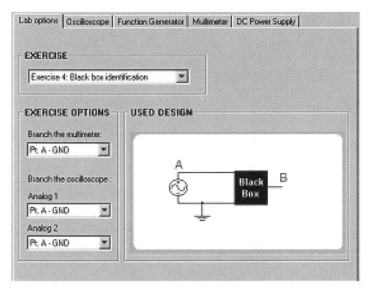

Figure 5.20   Exercise 4.

3. When you break a DC current through a solenoid, a high voltage will be generated across it. In this exercise the voltage in test point B in Figure 5.15 will only be slightly higher than the source voltage (5 V)—why? Check the trigger settings if you do not see a trace similar to that of Figure 5.15.

4. Find out what is in the black box in Figure 5.20. It is a network of R, L, or C, but what is the configuration. Do the measurements you need using the oscilloscope and the multimeter and draw the circuit diagram with component values specified.

5. Set the function generator to output a sinc wave 8 V_{pp}. Vary the frequency and measure $U_{A\text{-GND}}$ and $U_{C\text{-GND}}$ in Figure 5.21 as well as the phase relationship between them for the frequencies

$$f = f_0[0.1; 0.2; 0.4; 0.8; 1.2; 4.8; 10] \qquad (5.1)$$

Then draw the Bode plot. What is the bandwidth of the filter?

5.11.2 Transducer Laboratory Assignment 1: Line Scan Camera Experiments

Laboratory Assignment

• Design and implement a line scan camera using two signal sources and a camera box consisting of a linear sensor array TSL1401 mounted in the focal plane of a wide-angle TV camera lens. The TSL1401 linear sensor

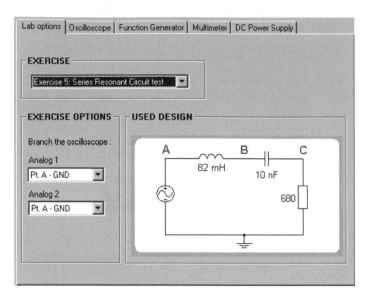

Figure 5.21 Exercise 5.

array consists of a 128×1 array of photodiodes and associated charge amplifier circuitry. Operation is simplified by internal logic requiring only a serial input signal and a clock. There are two function generators provided to generate those signals. Light energy striking a pixel generates a photo current, which is then integrated. The amount of charge accumulated at each pixel is directly proportional to the light intensity on that pixel and the integrating time. The integrating time defines the exposure.

• Check the performance of the camera.

There are four preparatory questions concerning the linear sensor array and its data sheet:

1. The data sheet can be downloaded from the Texas Advanced Optoelectronic Solutions website, http://www.taosinc.com/pdf/tslw1401.pdf. What are the recommended operating conditions for the input voltage (V_I), high-level input voltage (V_{IH}), and low-level input voltage (V_{IL})?
2. What is the voltage output from the sensor (white, average over 128 pixels, $V_{DD} = 5$ V)?
3. The exposure is determined by the time period of the SI pulse. What are the appropriate setup time $t_{su(SI)}$ and hold time $t_{h(SI)}$ as defined in Figure 5.22? Appropriate exposure with respect to light available and aperture set is approximately $\frac{1}{30}$ sec and then a clock frequency of 10 kHz

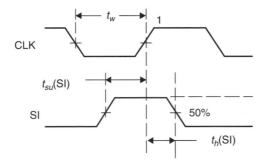

Figure 5.22   Part of timing diagram for TSL 1401.

will do. The sensor will work only if the transitions of the SI pulse and the first pulse of the clock (CLK) burst have a useful time relationship. Select a useful pulse width for the SI pulse considering the frequency selected for the pulses in the CLK burst. Also see note 1 below the table Recommended Operating Conditions in the data sheet.

4. The function generators in the remote laboratory can generate arbitrary waveforms. Only 500 samples of the generator arbitrary waveform memory are used. Use those to define one period of the SI signal. It should be a short pulse followed by a long pulse separation, that is, a few 1's and then 0's. At startup the memory is zeroed. How many 1's are appropriate?

Exercises

1. Design the camera using the camera box and the two function generators. The camera box has three inputs, V_{DD}, SI, and CLK, and an analog video output AO. V_{DD} is set to 5 V. Use the arbitrary waveform feature of the function generator A to generate the SI pulse. Function generator B is used to generate the CLK burst. At least 128 clock pulses are required to output one video line. Use the oscilloscope to check that the phase relationship and the voltage levels of the SI pulse and the first pulse of the CLK burst comply with the specification in the data sheet. If the voltage levels are not appropriate, then you are not allowed to connect the generators to the camera. If the phase relationship is wrong, the camera will not work.

2. Check the camera performance. Turn off the ceiling light in the laboratory. View one of the fluorescent lamps with the camera. How wide is the view angle? How many pixels are white when you view the three lamps one by one? Would it be possible to use these figures to roughly estimate the distance to the lamps?

3. By changing the exposure, you can change the camera sensitivity. Try to change the SI pulse frequency in the range 20–60 Hz and notice the change in sensitivity. This feature could be used to automatically control the exposure.

5.11.3 Transducer Laboratory Assignment 2: Design a Range Finder

Laboratory Assignment

- Design and implement a range finder using ultrasonic transducers.
- Check the performance of the range finder.

Ultrasound refers to sound waves at frequencies higher than the range of the human ear, that is, at frequencies greater than about 18 kHz. Ultrasonic waves obey the same basic laws of wave motion as lower frequency sound waves, but diffraction or bending around an obstacle of given dimensions is correspondingly reduced. It is therefore easier to detect and focus a beam of ultrasound. The theory for ultrasonic pulse echo meters can be found in some books on measurement systems. There are two ultrasonic transducers, a microphone and a load speaker attached to the fixture. These are used to find the distance to an obstacle and two walls; see the outline of the room in Figure 5.12. The preamplifier next to the microphone and shown in Figure 5.23 is required to feed a long cable. The other end of the cable is connected to a second amplifier, shown in Figure 5.24. The output of the second amplifier can be connected to the oscilloscope. The resonant frequency for the transducers is 40 ± 1 kHz and the maximum input voltage allowed for the load speaker is 10 V_{rms}.

Preparatory Questions

1. What is the maximum output voltage of the function generators?
2. The distance to the plank is less than 1.5 m. What is the time required for a sound burst to travel from the load speaker to the obstacle and back to the microphone if the distance to the obstacle is 1.5 m? The speed of sound is approximately 340 m/sec in the laboratory.

Exercises

1. Design and implement the range finder. Use function generator B to generate sine bursts to the load speaker and generator A to trigger those bursts. The frequency of the bursts must not be too high. A burst must travel from the load speaker to an obstacle at maximum distance allowed and back to the microphone before the next burst is fired. Use the oscilloscope as an indicator.
2. What is the distance to H1, H2, and H3 in Figure 5.25?

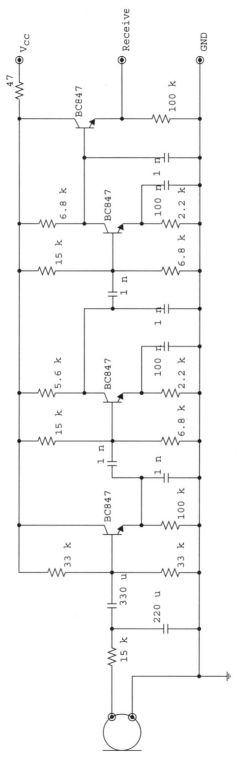

Figure 5.23 Microphone and preamplifier.

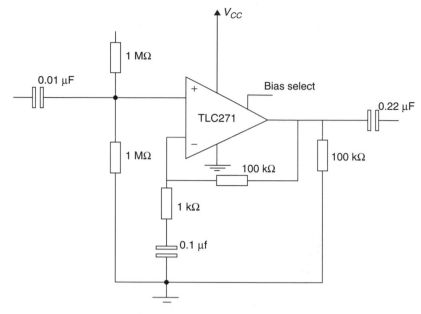

Figure 5.24   Microphone amplifier.

3. The measurement range depends on the strength of the sound pulse from the load speaker. What is the lowest useful input voltage to the load speaker?
4. The performance also depends on the number of periods in the burst. What is the appropriate number?
5. Is the performance degraded if square waves are used in the burst?
6. What is the distance from the load speaker to the microphone?

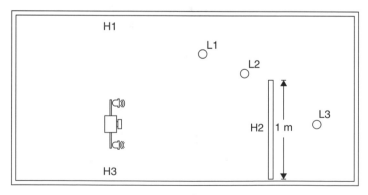

Figure 5.25   Outline of arrangements: L1, L2, and L3 are lamps; H1 and H2 are walls; and H2 is an obstacle (plank).

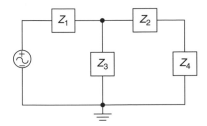

Figure 5.26 Simplified circuit diagram.

5.12 SOLUTIONS TO EXPERIMENTS AND LABORATORY ASSIGNMENTS

5.12.1 Electrical Experiments

Preparatory Questions

1. Use the designations in the simplified circuit diagram in Figure 5.26. All impedances are in ohms:

$$Z_1 = 2.7 \times 10^3 \tag{5.2}$$

$$Z_2 = Z_1 \tag{5.3}$$

$$Z_R = 10 \times 10^3 \tag{5.4}$$

$$Z_C = \frac{1}{jwC} = \frac{1}{j2\pi \times 10^3 \times (56 \times 10^{-9})} \approx -j2.842 \times 10^3 \tag{5.5}$$

$$Z_3 = \frac{Z_R Z_C}{Z_R + Z_C} \approx 747.4 - j2.630 \times 10^3 \tag{5.6}$$

$$Z_4 = Z_3 \tag{5.7}$$

$$Z_5 = Z_3 // (Z_2 + Z_4) = \frac{Z_3(Z_2 + Z_4)}{Z_3 + Z_2 + Z_4} \approx 879.7 - j1.527 \times 10^3 \tag{5.8}$$

$$U_1 = \frac{Z_5}{Z_1 + Z_5} U \approx (0.3618 - j0.2722) U \tag{5.9}$$

$$|U_1| \approx 0.4528 \cdot U \qquad \angle U_1 \approx -36.9° \tag{5.10}$$

$$U_2 = \frac{Z_4}{Z_2 + Z_4} U_1 \approx (0.5049 - J0.3777) U_1 \tag{5.11}$$

$$|U_2| \approx 0.6305 |U_1| \qquad \angle U_2 \approx -36.8° \text{ referenced to } U_1 \tag{5.12}$$

$$U_3 = \frac{Z_2}{Z_2 + Z_4} U_1 \approx (0.4951 + j0.3777) U_1 \tag{5.13}$$

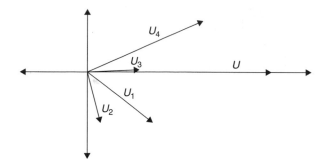

Figure 5.27 Calculated phasor diagram.

$$|U_3| \approx 0.6227|U_1| \qquad \angle U_3 \approx 37.3° \text{ referenced to } U_1 \qquad (5.14)$$

$$U_4 = \frac{Z_1}{Z_1 + Z_5} U \approx (0.6382 + j0.2722)U_1 \qquad (5.15)$$

$$|U_4| \approx 0.6939U \qquad \angle U_4 \approx 23.1° \qquad (5.16)$$

A phasor diagram is shown in Figure 5.27.

2. $f_0 = \dfrac{1}{2\pi\sqrt{LC}} = \dfrac{1}{2\pi\sqrt{(82 \times 10^{-3})(10 \times 10^{-9})}} \approx 5558 \text{ Hz} \qquad (5.17)$

Figure 5.28 shows the simulated Bode plot.

Exercises

1. $U = 4$ V DC. The results of the measurements are shown in Table 5.2.
2. $|U| = 8$ V_{pp} AC. The results of the measurements are shown in Table 5.3. Calculate the phasors in rms volts from the measured values in Table 5.4:

$$U = 2.80 \qquad (5.18)$$

$$U_1 = 1.0177 - j0.7394 \qquad (5.19)$$

$$U_2 = 0.2407 - j0.7408 \qquad (5.20)$$

$$U_3 = U_1 - U_2 = 0.777 - j0.001 \qquad (5.21)$$

$$U_4 = U - U_1 = 1.7822 - j0.7394 \qquad (5.22)$$

A phasor diagram is shown in Figure 5.29. In Table 5.5 the length of the calculated phasors and the multimeter results are compared.

3. You must change the trigger settings for the oscilloscope to get the trace shown in Figure 5.15. The oscilloscope settings together with a somewhat

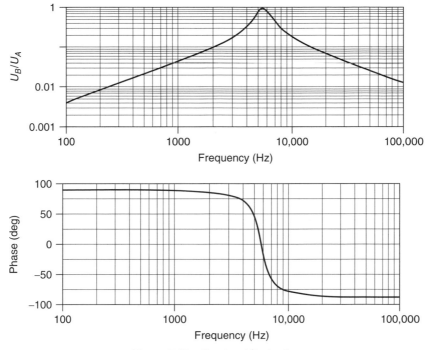

Figure 5.28 Simulated Bode plot.

Table 5.2   DC Measurements with Multimeter (V)

U_1	U_2	U_3	$U_1 - U_3 - U_2$
2.69	2.12	0.57	$2.69 - 0.57 - 2.12 = 0$

Table 5.3   AC Measurements with Multimeter (V)

$|U_1|$	$|U_2|$	$|U_3|$	$|U_1| - |U_3| - |U_2|$
1.258	0.779	0.790	$1.258 - 0.79 - 0.779 = -0.311$

Table 5.4   Phasor Measurements with Oscilloscope

Parameter	U	U_1	U_2
Amplitude (V_{rms})	2.80	1.258	0.779
Phase (deg)	0	36	72

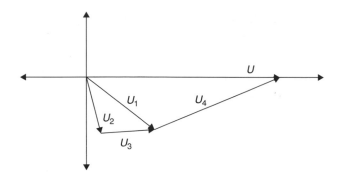

Figure 5.29 Measured phasor diagram.

Table 5.5   Length of Phasors Calculated Compared with Measurement

Measured in Diagram		Measured with Multimeter	
$|U_3|$	$|U_4|$	$|U_3|$	$|U_4|$
0.777	1.93	0.790	1.937

expanded sweep are shown in Figure 5.30. A probe with 10 times attenuation is used. The current is not broken. It will pass through the oscilloscope when the switch is opened. A PSPICE circuit with the oscilloscope included is shown in Figure 5.31. The resistance of the solenoid is measured. With the component values selected, a simulation in PSPICE will give approximately the same outcome as in Figure 5.30.

4. If you set a DC offset on the input, it will be blocked. Thus there must be a capacitor between the test points A and B (see Figure 5.20). Measure and calculate the amplitude ratio and phase shift for some frequencies.

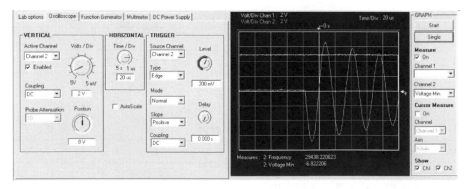

Figure 5.30 Oscilloscope settings for Exercise 3. Bottom trace is transient on channel 2.

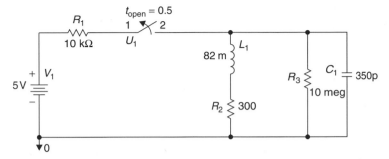

Figure 5.31   PSPICE simulation of Exercise 3.

Data for some selected frequencies are shown in Table 5.6. A voltage ratio diagram is shown in Figure 5.32. It tells you that there is a first-order high-pass filter in the black box ($f_0 \approx 5600$ Hz). The configuration is shown in Figure 5.33:

$$RC = \frac{1}{2\pi \times 5600} = 2.8 \times 10^{-5}$$

Set the function generator to output a low voltage of low frequency. Then the capacitor will block the function generator output and you can use the multimeter to measure the resistance. The correct component values are 2.7 kΩ and 10 nF.

5. Table 5.7 shows the measured data and Figure 5.34 shows the corresponding Bode plot. The bandwidth is 2200 Hz.

Table 5.6   Bode Plot Data

Frequency	U_{out}	U_{in}	Phase (deg)	$20 \log(U_{out}/U_{in})$
500	0.248	2.825	84	−21.13134
1000	0.4667	2.83	80	−15.65497
2000	0.892	2.826	70.5	−10.01615
4000	1.536	2.795	55	−5.19981
5000	1.72	2.71	48.5	−3.94882
5500	1.79	2.78	46	−3.82384
5600	**1.81**	**2.77**	**45**	**−3.69602**
5700	1.8	2.76	44.5	−3.71273
6000	1.89	2.72	42.5	−3.16214
8000	2.18	2.87	36	−2.38851
16000	2.58	2.88	20	−0.95546
32000	2.65	2.8	10.5	−0.47824
64000	2.64	2.75	5.5	−0.35458

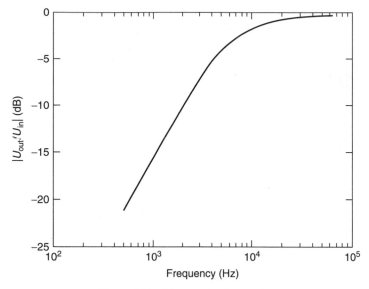

Figure 5.32   Measured voltage ratio.

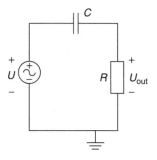

Figure 5.33   Circuit in black box.

Table 5.7   Measured Data

Frequency	U_{out}	U_{in}	$20 \log(U_{out}/U_{in})$
555.79	0.065484	2.900655	−32.92722
1,111.6	0.137828	2.833598	−26.26002
2,223.2	0.311597	2.813322	−19.11252
4,446.3	1.159646	2.758263	−7.52620
5,557.9	1.630321	2.715144	−4.43039
6,947.4	1.055216	2.774105	−8.39563
1,111.6	0.40973	2.821713	−16.76030
22,232	0.156833	2.804472	−25.04827
44,463	0.034546	2.83858	−38.29407
55,579	0.015272	2.814167	−45.30908

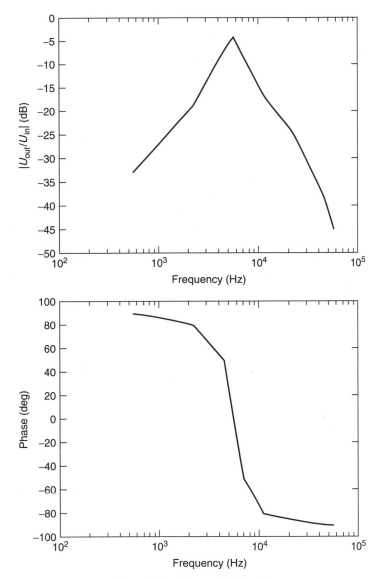

Figure 5.34 Measured Bode plot.

5.12.2 Transducer Laboratory Assignment 1: Line Scan Camera Experiments

Preparatory Questions

1. The recommended operating conditions are
   Input voltage is $0 < V_I < V_{DD}$.

High-level input voltage is $V_{DD} \times 0.7 < V_{IH} < V_{DD}$.

Low-level input voltage is $0 < V_{IL} < V_{DD} \times 0.3$.

2. This analog output voltage is typically 2 V.

3. The setup time $t_{su(SI)}$ is a minimum of zero, which means that SI goes high before the rising edge of the clock pulse. The hold time $t_{h(SI)}$ is a minimum of 20 nsec. Due to note 1, SI must go low before the rising edge of the next clock pulse. This means that the SI pulse can be approximately the same length as the clock pulse, that is, 50 μsec.

4. A 1 and 499 0's will generate a pulse width for the SI pulse of $\frac{1}{500} \times \frac{1}{30} = 67$ μsec.

Exercises

1. Start the transducer lab client program. The `Sync` output of function generator A will then be connected to the `Ext Trig` input of function generator B. Only generator B can generate a burst. Thus function generator A must be used to generate the SI signal and B to generate the CLK burst. At startup some settings of the instruments are already made so you need not change them.

   a. Start to generate the SI signal. The settings of function generator A are shown in Figure 5.35. Change the waveform text box to `User defined` and the frequency to 30 Hz. You can either turn the frequency control knob using the mouse or write 30 in the text box beneath the knob and press `Enter`.

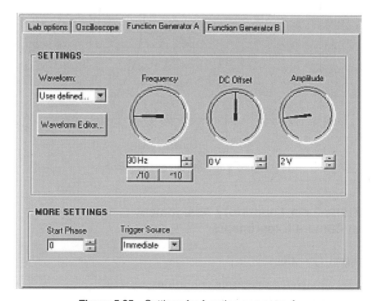

Figure 5.35 Settings for function generator A.

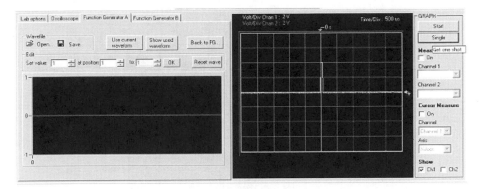

Figure 5.36 Waveform editor.

b. Press the `Waveform Editor` key. A 1 in the first position in the waveform memory is default, as is shown in Figure 5.36. Then press the keys `OK` and `Use current waveform`. Press the key `Single` in the upper right corner of the client program window to get a trace on the oscilloscope display. Clear the `Show Ch 2` box. You should now see the SI pulse in green on the oscilloscope display, as shown in Figure 5.36.

c. Show the oscilloscope panel. Move the trace downwards by writing −6 in the text box below position control and press the key `Single` again.

d. Show the panel of the function generator B. Select waveform `Square`, frequency 10 kHz, DC offset 1 V, and amplitude 1 V; check the `Burst Mode` box, and set the count to 130. Check the `Show Ch 2` box and press the key `Single` again. You should now see the CLK burst together with the SI signal on the oscilloscope display, as shown in Figure 5.37.

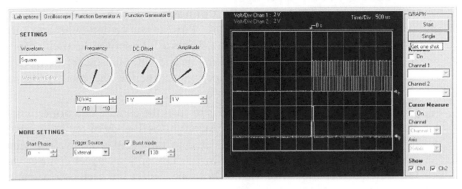

Figure 5.37 Setting for function generator B. Bottom trace is SI pulse on channel 1. Top trace is showing start of burst on channel 2.

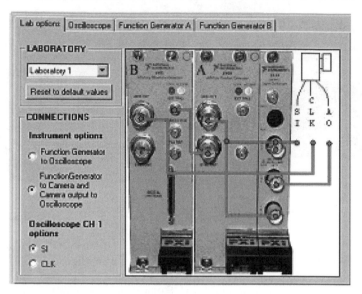

Figure 5.38   Laboratory options panel. Function generators are connected to camera.

e. Show the lab options panel. Select the option `Function Generator to Camera` and `Camera output to Oscilloscope`. Then the function generators will be connected to the camera and the camera output to channel 1 of the oscilloscope, as shown in Figure 5.38, if all the settings of the function generators are correct. You will get an error message if some voltage setting is wrong.

f. Show the oscilloscope panel again. Turn the `Time/Div` knob to 2 msec or write 0.002 in the text box beneath the knob and press `Enter`. Turn the `Delay` knob to approximately 6 msec. Press the `Stream video` key in the upper right part of the client window to receive the web camera video and press the `Connect` key in the left upper corner to log on to the transducer fixture. Put out the ceiling lamp and turn on the lamp closest to the fixture. Figure 5.39 shows the oscilloscope display when the camera is looking at the lamp closest to the fixture. Use the two buttons in the upper left corner of the client program window to turn the camera fixture.

2. One way to measure the view angle is to turn the camera and notice when the lamp disappears from the video signal at both ends of the image line. The direction of the camera is indicated in the client window. The view angle is somewhat less than 90°. On the lab options panel connect the CLK burst to channel 1 of the oscilloscope. Select the oscilloscope panel and change the time scale to zoom, as in Figure 5.40. You might have to adjust the trig delay. The number of white pixels is 11, 8, and 7.

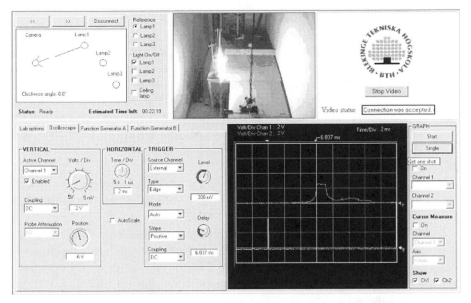

Figure 5.39 Camera is looking at lamp 1. Bottom trace is SI pulse on channel 1. Top trace is video signal on channel 2.

5.12.3 Transducer Laboratory Assignment 2: Design a Range Finder

Preparatory Questions

1. The maximum output voltage of the NI 5411 function generator is ±10 V into a high-impedance load.
2. The travel time is $2 \times 1.5/340$ sec = 8.8 msec.

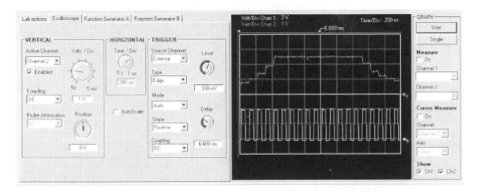

Figure 5.40 Expanded time scale. Top trace is video signal on channel 2. In time scale selected individual pixels can be identified. Bottom trace is CLK signal on channel 1.

Exercises

1. Use the radar principle to design the range finder. A burst containing some 40-kHz periods is required to make the load speaker oscillate and output a sound pulse. The microphone must then receive the echo before the load speaker transmits the next burst. If the silent period between the bursts is approximately 10 msec, the maximum range will be $340 \times 0.01/2 = 1.7$ m. Let us try to design a meter with this maximum range. Use function generator A to set the delay between successive bursts to the load speaker and function generator B to generate the bursts.

   a. Use function generator A to generate a square-wave trigger signal to generator B with frequency 100 Hz, DC offset 1 V, and amplitude 1 V. The settings are shown in Figure 5.41.

   b. Use the function generator B to generate a burst of 10 periods of a sine wave with frequency 40 kHz, DC offset 0, and amplitude 2.47 V. The settings are shown in Figure 5.42.

   c. Change the settings of the oscilloscope to channel 2 vertical 5 V/div, horizontal 1 msec/div, and trigger delay 5 msec. Then press the key Single to get oscilloscope traces like those in Figure 5.43.

2. Select laboratory exercise 2 using the lab options panel and the option Function Generator to Ultrasound and Ultrasound back to Oscilloscope. Then the function generator B will be connected to the load speaker and the microphone via two amplifiers to channel 2 of the

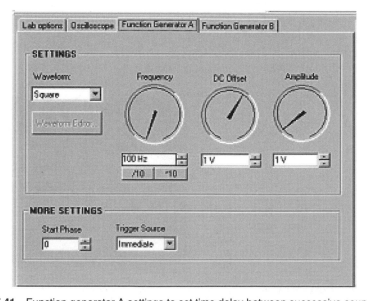

Figure 5.41 Function generator A settings to set time delay between successive sound bursts.

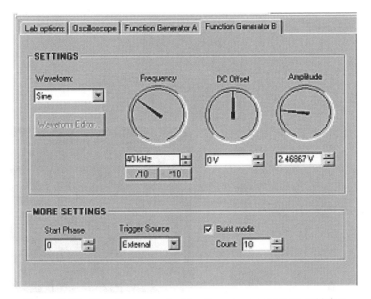

Figure 5.42   Function generator B settings to generate sound burst.

oscilloscope, as shown in Figure 5.44, if all the settings of the function generators are correct. In Figure 5.45 the range finder is directed towards H3. Turn the fixture and measure the distances H1, H2, and H3. They should be 104, 129, and 56 cm, respectively.

3. The minimum useful amplitude of the drive waveform depends on the performance of the detector and the amplifiers.

4. Only a few periods are needed.

5. It is not significantly degraded.

6. The distance is approximately 15 cm.

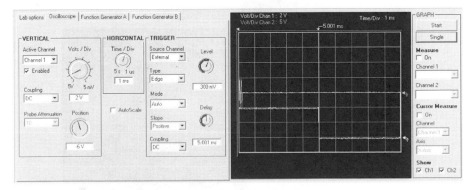

Figure 5.43   Oscilloscope settings to show range finder output. Top trace is burst on channel 2 and bottom trace is 100-Hz square wave on channel 1.

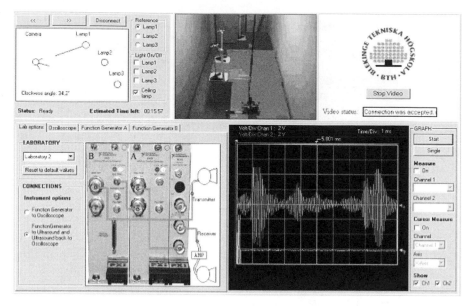

Figure 5.44 Output from range finder when it is directed toward plank. Top trace is echo on channel 2. Bottom trace is burst output to load speaker.

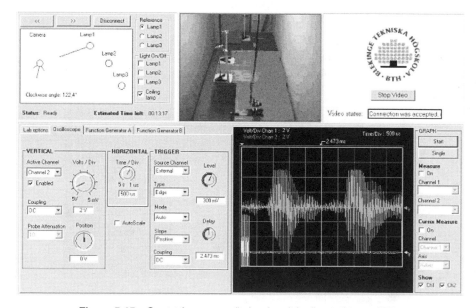

Figure 5.45 Ouptut from range finde when it is directed toward H3.

5.13 CONCLUSION

Many experiments in electrical engineering education have no physical sensations and can be conducted remotely over the Internet around the clock and without video transmission or other means requiring high transfer bandwidth. In experiments with short time constants, several students can share the same remote hardware.

It is also possible to manage experiments with long time constants and physical sensations; only one client can then control the experiments, and video and/or sound transmission is required. At BTH, we have combined the two types of experiments in two laboratory assignments and used them in an ordinary course in December 2001. Only one student can control the experiment from any one location, but other students from other locations can observe the physical sensations and perform parts of the experiments. This approach has not yet been evaluated, although some students say they prefer to work in the local laboratory in the conventional way. Others appreciate the opportunity to do the experiments when they choose and using the time they need.

REFERENCES

J. Bentley, *Principles of Measurement Systems*, 3rd ed., Longman, Singapore, 1995.

D. Comer, *Internetworking with TCP/IP*, 3rd ed., Vol. I, Prentice-Hall, Englewood Cliffs, New Jersey, 1995.

D. Cunningham, *Circuit Analysis*, 2nd ed., Houghton Mifflin, Boston, 1995.

R. Goody, *OrCAD Pspice for Windows*, 3rd ed., Vol. II, Prentice-Hall, Englewood Cliffs, New Jersey, 2001.

D. Johnson, *Electric Circuit Analysis*, 3rd ed., Prentice-Hall, Englewood Cliffs, New Jersey, 1997.

J. Keown, *OrCAD Pspice and Circuit Analysis*, 4th ed., Prentice-Hall, Englewood Cliffs, New Jersey, 2001.

B. Kernighan, *The C Programming Language*, Prentice-Hall, Englewood Cliffs, New Jersey, 1978.

J. Nilsson, *Introductory Circuits for Electrical and Computer Engineering*, Prentice-Hall, Englewood Cliffs, New Jersey, 2002.

6

REMOTE LABORATORY: BRINGING STUDENTS UP CLOSE TO SEMICONDUCTOR DEVICES

A. Söderlund, F. Ingvarson, P. Lundgren, and K. Jeppson

Chalmers University of Technology, S-412 96 Göteborg, Sweden

6.1 INTRODUCTION

From an educational point of view, the term *remote laboratory* may actually be misleading when referring to techniques of doing measurements on devices connected on-line in one part of the world from a computer located somewhere else. By establishing an electronic link between the student's desktop and a semiconductor device, we bring students in close contact with the device under study, eliminating the need to learn about complicated measurement instruments or of having to rely on second-hand information from books or teachers. No locked laboratory doors stand in the way of gaining knowledge and the device is immediately available for first-hand study for each and every student. So even if the device and student are physically remote, the experimental intimacy provided by the remote laboratory is a fundamental and characteristic quality of this teaching tool. If you like, the acronym for our endeavors in this

Lab on the Web: Running Real Electronics Experiments via the Internet
Edited by Tor A. Fjeldly and Michael S. Shur
ISBN 0-471-41375-5   Copyright © 2003 John Wiley & Sons, Inc.

field at Chalmers University of Technology, I-Lab, could be read as either Internet laboratory or intimate laboratory.

The Internet is an ideal medium for remote instruction purposes (see Shen et al., 1999; Fjeldly et al., 2002; del Alamo et al., 2002; Söderlund et al., 2002; Geoffroy et al., 2001). Its protocol standards make data communication and graphical user interfaces easy to implement. National Instruments (see references) offers software packages that allow us to use the Internet for remote operation of lab instrumentation. Their Internet developers toolkit makes virtual instrument front panels viewable from standard web browsers by converting the front panels into images.

The concept of virtual instrumentation is to create a more powerful, flexible, and cost-effective instrumentation system built around a computer using software to control the instrument setup and provide an intuitive and user-friendly interface. A virtual instrument can easily export and share its data and information with other applications.

We have used these techniques as a replacement for traditional laboratory exercises in a compulsory course on microelectronic devices for 350 second-year undergraduate students at Chalmers. In this course, I-Lab is used for characterization of semiconductor diodes. In the following, we will discuss both the technical aspects of I-Lab as well as the issues of how to integrate this learning tool into engineering education to harness its full potential.

6.2 THE I-LAB SYSTEM

The I-Lab system consists of high-performance measurement instruments for electrical DC and low-frequency AC characterization of two-terminal devices. We have chosen to use these rather expensive measurement instruments to allow accurate and straightforward characterization of semiconductor devices. Our experience from doing research on semiconductor physics and devices, where we use such instrumentation everyday, is that these instruments are easy to use and offer high reliability and high accuracy.

In I-Lab, the instruments and the set of devices to be characterized are all connected to a switch unit. The switch makes it possible through computer control to choose which instrument is to be physically connected to any one of the devices connected to the switch. By using this approach, it is straightforward to perform both DC and AC characterization of several different devices without having to manually switch between instruments and devices. This is an essential property of a remote laboratory. Figure 6.1 shows the I-Lab setup where we use a Keithley 236 source measure unit (SMU) for DC characterization and a Hewlett-Packard 4284A *LCR* meter for low-frequency AC characterization of semiconductor diodes. By using an Agilent E5250A switch, one can decide whether the SMU or the *LCR* meter is to be physically connected to the selected one of the five available diodes.

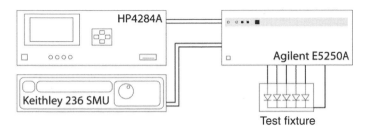

Figure 6.1   Agilent E5250A switch allows HP4284A *LCR* meter or Keithley 236 SMU to be physically connected to one of five semiconductor diodes in test fixture.

The Keithley 236 SMU offers $I(V)$ [and $V(I)$] measurement possibilities from sub-picoampere current levels up to 100 mA with high accuracy. This is very important for our purposes where DC characterization of semiconductor diodes requires current measurements covering 10–15 orders of magnitude. For accurate low-level measurements, a fully guarded approach with triaxial cables is employed. This allows the guard of the SMU to be extended all the way to the test fixture, thus reducing the influence of charging currents and electromagnetic interference (EMI) during measurement of very low currents. In addition, the test fixture is designed to be an effective shielding box. For high current levels, remote sensing can be utilized for removing the effects of parasitic series resistances.

In I-Lab, the SMU is programmed to perform a voltage sweep and measure the current. Measurement data are stored in an internal memory during the sweep and are retrieved by the computer upon completion of the sweep. To have a reasonable measurement time for a sweep, instrument settings like integration time, degree of filtering, delay time, and so on, can be adjusted for minimizing the time required for a measurement for any given measurement accuracy.

The HP4284A *LCR* meter is used for measuring the impedance, in the frequency range of 20 Hz–1 MHz, of a two-terminal device for different bias points. Capacitance, inductance, and resistance values are then determined by the instrument by selecting one of several equivalent circuit models (C_{series}–R_{series}, $C_{parallel}$–$R_{parallel}$, L_{series}–R_{series}, etc.) that best represents the device being characterized. Capacitances in the sub-0.1-fF range can be measured using the HP4284A.

The Agilent E5250A switch is equipped with a 10×12 matrix switch card that can be programmed to connect any of the 10 input connectors to any of the 12 output connectors. Low-level measurements required for a full diode DC characterization requires that the switch offer a very high signal-to-ground isolation resistance. The low-leakage ports of the switch card have a very high isolation resistance of about 10^{13} Ω, thus enabling accurate low-level measurements. The *LCR* meter is connected to the $C(V)$ input port on the switch unit.

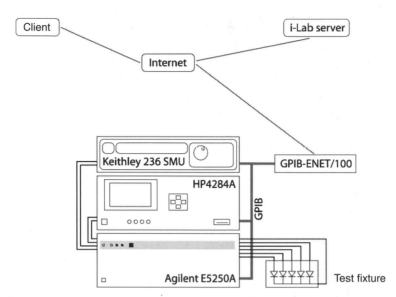

Figure 6.2 I-Lab system.

All measurement instruments are interconnected through a GPIB bus, which in turn is connected to the Internet using a National Instruments GPIB-ENET/ 100 GPIB–ethernet bridge. The I-Lab server controlling all instrumentation is also connected to the Internet. By using this approach, a very flexible system is obtained allowing the server to be located anywhere on the Internet. At our laboratory, we have the I-Lab server running on a computer, which houses our traditional web server. This computer is not associated with the measurement laboratory where the instrumentation is located. The entire I-Lab system is shown in Figure 6.2. A client communicates with the I-Lab server via the Internet and the server communicates with the instrumentation over the GPIB–ethernet bridge. The I-Lab server is implemented in LabVIEW 6i from National Instruments, running on a Linux server. LabVIEW 6i has a web server available, which we have utilized for the communication with clients. The part of the I-Lab server controlling the measurement instruments is an in-house solution implemented in LabVIEW.

As evident from Figure 6.2, I-Lab is based on a client–server structure. A client can request measurements to be performed by the I-Lab server through one of two interfaces; a regular web page accessed through a standard web browser and a LabVIEW Player application. The regular web page is the simplest interface. On the web page, the user can choose between performing $I(V)$ or $C(V)$ measurements by clicking on the appropriate link. After choosing type of measurement, a new web page is shown (see Figure 6.3). This page contains dialogue boxes where users enter information about the $I(V)$ or $C(V)$ measurement that they want to perform. The user can set the number of data

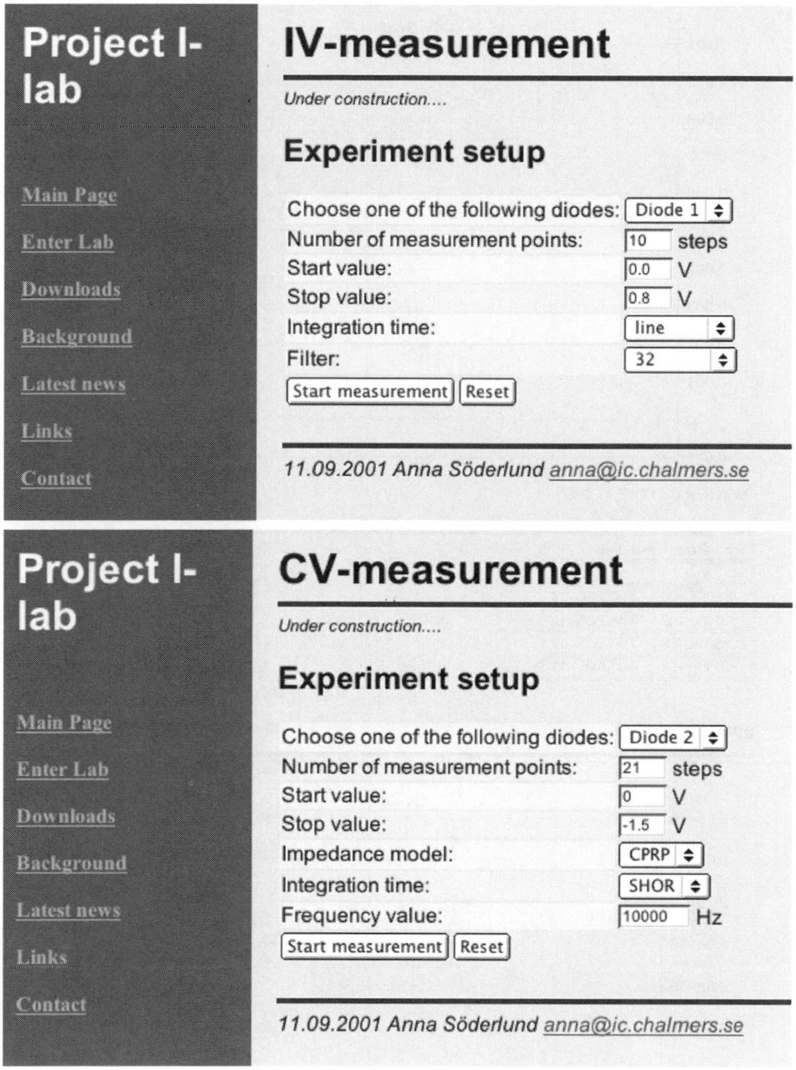

Project I-lab

Main Page

Enter Lab

Downloads

Background

Latest news

Links

Contact

IV-measurement

Under construction....

Experiment setup

Choose one of the following diodes: [Diode 1 ⬍]
Number of measurement points: [10] steps
Start value: [0.0] V
Stop value: [0.8] V
Integration time: [line ⬍]
Filter: [32 ⬍]
[Start measurement] [Reset]

11.09.2001 Anna Söderlund anna@ic.chalmers.se

Project I-lab

Main Page

Enter Lab

Downloads

Background

Latest news

Links

Contact

CV-measurement

Under construction....

Experiment setup

Choose one of the following diodes: [Diode 2 ⬍]
Number of measurement points: [21] steps
Start value: [0] V
Stop value: [-1.5] V
Impedance model: [CPRP ⬍]
Integration time: [SHOR ⬍]
Frequency value: [10000] Hz
[Start measurement] [Reset]

11.09.2001 Anna Söderlund anna@ic.chalmers.se

Figure 6.3 Web page client interfaces for $I(V)$ and $C(V)$ measurements.

points, the start and stop values, the integration time, the degree of filtering, the impedance model and frequency, and so on. When users are satisfied with the settings, they can send a request for a measurement by pressing the Start measurement button on the web page. The request is then received by the server and put into a queue managed by the I-Lab web server. All queued requests are processed in a first-in, first-served fashion. After completion of a measurement, I-Lab generates a new web page containing a graph and a table with the measurement data (see Figure 6.4). This new web page is opened on

IV measurement results

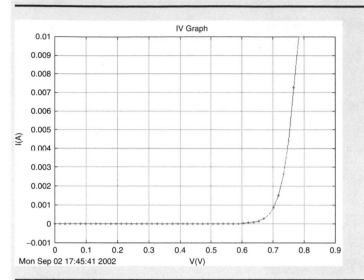

Click here to save data

V	I
0.00000E+00	−4.90000E−13
1.63265E−02	8.45000E−12
3.26531E−02	1.87700E−11
4.89796E−02	3.23000E−11

CV measurement results

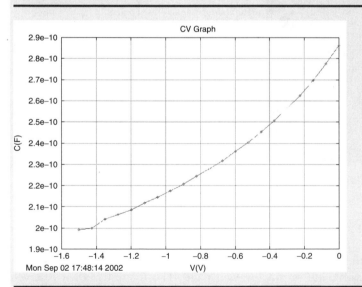

Click here to save data

V	C
0.00000E+00	2.86180E−10
−7.50000E−02	2.77523E−10
−1.50000E−01	2.69604E−10

Figure 6.4 Web page interface for returned results from I-Lab server.

the client side as a pop-up window. The user has the option to save the measurement data to disk for further processing in other programs such as Matlab and Excel.

The input parameters in the web page interface have a limited range of valid values for preventing input values to be sent to the server that can lead to experiment failure. The user receives an error message if that is the case and new values must then be entered. If an error should occur that could lead to malfunction of the server, and hence affect the communication between the server and the client or between the server and the measurement equipment, the server sends an e-mail to inform the administrator, who can then correct whatever has failed.

I-Lab can also be accessed through a client application implemented as a LabVIEW virtual instrument (or VI). To use the client VI, LabVIEW or the free-of-charge LabVIEW Player must be installed on the client computer. LabVIEW Player is available for download from the National Instruments website. To access I-Lab, the user must download the client VI from the I-Lab website and then open the VI in LabVIEW or LabVIEW Player. Figure 6.5 shows the client VI interface. LabVIEW offers much more straightforward means for creating easy-to-use user interfaces compared to the web page solution. For instance, graphs in LabVIEW VIs are easily configured by the user during runtime; changing from linear to log scales and zooming in and out are tasks readily performed. However, a drawback with the LabVIEW Player approach is that the Player application is about 19 Mbytes, requiring a high-bandwidth Internet connection to obtain a reasonable download time. A sec-

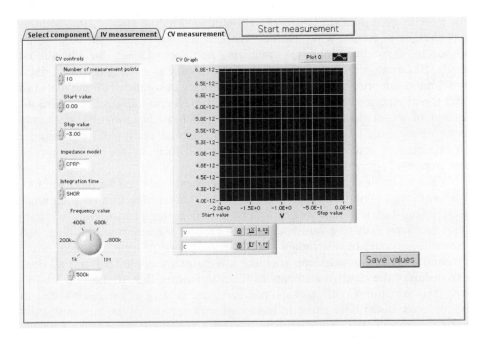

Figure 6.5   LabVIEW Player running I-Lab client application.

ond drawback is that Player is not available for Linux, which is a popular operating system, at the time when this chapter was written. We have tested the system from off-campus using a standard 56-kbps modem. According to our tests, the time needed to access the website and to open the client web page interface is about 30 sec or less. It takes about 20 sec or less for the system to perform a complete measurement with 10 data points, from sending the request to receiving and plotting the measured data.

6.3 USING I-LAB AS A LEARNING TOOL

A traditional laboratory exercise concerning device characterization in electrical engineering at Chalmers is often defined by a set of measurement and analysis tasks that are to be performed using a suggested experimental setup. This either is to be replicated by the students or is already in place and ready for use. The limited time available in the traditional laboratory often leads to streamlining of the experimental tasks in order for the students to be able to cover as much ground as possible in the time given. The situation is far from a realistic experimental situation since it does not allow sufficient time for the students to learn from their mistakes or for them to take a creative part in experimental problem solving. One could argue that it is a waste of equipment and resources when the students use instruments only to gather data, without being given a chance to reflect in depth on the measurement procedure. If the data are important, why not just hand them over to the students right away? And if the experimental skill is to be trained, why not let the students deal with the experimental problems without streamlining? Finding answers to questions like these is, of course, what learning is about, and we believe that I-Lab is an important alternative tool when optimizing resources to reach our educational goals. I-Lab can provide easy access to data, as easy as picking it from a database, but it can preserve the decisive sense of being in contact with the real stuff that a real experiment conveys and that has a profound influence on the motivation of the student. I-Lab cannot train practical experimental skills any better than a fixed lab setup with streamlined exercises, but it will free resources to expand such activities, since the activities that focus on the devices can be run in a more efficient manner.

When we implemented I-Lab, the lab assignment was open for modification, with enhanced opportunities for incorporating student reflection on data analysis. Since I-Lab is always available for the student, the previous limited time-frame for data gathering vanished, and the process of measuring, thinking, and remeasuring could be distributed over a longer period of time. Practical issues of connecting instruments and suchlike was naturally minimized and allowed expansion of the creative elements in the assignments. The traditional laboratory, which requires both in-lab supervision and control of the results, had to be replaced. Instead, we chose to have oral presentations of the lab results and gave the students the opportunity to contact a supervisor (by e-mail or otherwise) when need arose.

The students were given the freedom to independently organize their laboratory exercise and to determine, within some given constraints, which measurements to perform.

In the traditional laboratory setup, which we used for many years, we had a number of instrumentation setups available in the laboratory hall. The students often arrived poorly prepared, trying to figure out what to do during the laboratory session by reading the manual while progressing from task to task by frequently asking questions of the supervisor.

In its new form, the laboratory exercise was formulated as an open task, where we asked the students to come up with a number of parameters that they could argue were important for characterizing a semiconductor diode. They were then asked to tell us how these parameters could be determined through measurements and to actually determine all or some of these parameters for a randomly assigned diode by using the I-Lab system.

In the new form of the laboratory exercise, the students were given about two weeks to perform the task—each student having a partner so that they were working in groups of two. After the two weeks, the oral presentations took place. Each group was given 9 min to present their findings to a teacher and to four other groups. The tight time schedule required the students to come well prepared with their data and findings properly presented on up to four or five transparencies. All of the presentations were concentrated to two days with four presentation sessions running in parallel. Each session was chaired by a professor or lecturer who determined which presentations were to be regarded as excellent (by predefined, well-specified guidelines available to the students). This way, each teacher assessed 90 students.

Since this was a first-time experience, we did not know what to expect from the student presentations. We had some fears of complaints since the system had gone down a number of times, up to 24 hr during the first weekend. This is further described in the Technical Problems section below. On the other hand, most students appeared very relaxed, often showing self-confidence and pride in what they had achieved during the laboratory exercise. Compared to the reluctance sometimes expressed by the students toward the traditional organization of the laboratory, this was a positive experience and definitely to be preferred. Of the student presentations from all the sessions, less than 10 presentations were below standards—most were very good presentations. About 20% of the presentations were excellent and were accredited with bonus points.

After completion of the course, we received written review forms by the students, and the result of the evaluation is presented in the next section.

6.4 EVALUATION

During the last two terms, two different student groups have used I-Lab. First, a small group of final-year students was given an assignment to solve using the remote laboratory. Our purpose was to get useful feedback for further improvements of I-Lab before introducing it to the large class of 350 second-year students. The group gave us positive feedback and valuable suggestions that were used to improve the remote laboratory.

The students could choose whether they wanted to use LabVIEW Player or the regular web page as an interface to access the measurement equipment. Eighty-five percent of the students preferred the easy access offered by the web page interface. A reason for this could be that a majority of the students used their private computers from home, having only a slow connection to perform measurements. Therefore, they did not consider it worthwhile to download the LabVIEW Player application of 19 Mbytes, when an easier access alternative was offered. Another reason was that LabVIEW Player is not available for UNIX or Linux. UNIX is the operating system used by the university. Ninety percent of the students performed the measurements after school hours in the evening. None of the final-year students had to wait more than 2 min for their measurement results. The students also appreciated the possibility of being able to redo the measurement if they noticed that some values were questionable or appeared to be incorrect.

I-Lab was also used in the introductory microelectronic device course offered to second-year computer engineering and electrical engineering students, as mentioned above. Unlike the final-year students, the second-year students had not yet adopted a research-oriented attitude toward learning. Almost all of the students felt that the task attached to the remote laboratory was more demanding than for traditional laboratories, but they also felt that the task was much more instructive. In this group of students, only a minority performed the measurements from home, while most students used university computers. Still, 70% of the students performed their measurements in the evening, after school hours. Around half of the second-year students had to wait for more than 5 min because of server problems, but the other half of the students were very pleased with the short time needed to perform a measurement.

Afterward, a majority of the students in both groups were pleased with the remote laboratory, and they felt that it had more advantages than disadvantages. Among other things, they particularly pointed out the advantages of being able to access the laboratory at anytime, of being able to perform the measurements from home (which was very convenient), and of being able to redo their measurements if they noticed some questionable data.

We discovered that one of the disadvantages of performing the lab via the Internet was that most of the students in both groups missed having an instructor to consult with. About half of the students missed not having hands-on contact with the measurement equipment. The other half said that it usually takes a long time to figure out how the equipment works and that it is better to concentrate on the data analysis instead.

6.5 TECHNICAL PROBLEMS

With a small group of students, around 15, there were no problems for the LabVIEW server to hold that number of clients in queue, but for a larger

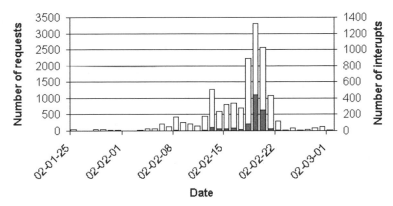

Figure 6.6 Number of measurement requests per day and number of requests that were interrupted (i.e., no client remained to receive data after measurements were completed).

group, around 350 students, this became a problem. The server went down a number of times and needed to be restarted. A temporary solution was to make the LabVIEW server automatically restart after a server crash.

We do not know yet why the system goes down. The queue system can easily handle a large number of requests. However, the queue can grow extremely large when a client experiences no immediate response and therefore sends multiple requests. The queue then grows rapidly since a request is only deleted from the queue after being properly processed by the server. A client can thus have several identical requests queued. On the night before the deadline, the number of requests exceeded 3400, which corresponds to more than 10 requests per student. No wonder that there were complaints about poor access to the server during the last 24 hr before presentations were due on February 21 and 22 (see Figure 6.6). Of all the submitted requests, a large number were "interrupted" since the clients were unavailable to receive data by the time that their measurements were completed. But they still occupied the measurement equipment. If each measurement request takes about 30 sec to handle, this corresponds to more than 24 hr of measurements each day. Therefore, a much better measurement discipline must be enforced by next year, since there is no reasonable way that the server could handle such a large number of requests within any reasonable amount of time.

6.6 DISCUSSION

We believe that using I-Lab to provide real laboratory experiments via the Internet is a situation that cannot be replaced by simulation software packages without losing important features. When replaced by simulation software or a database, the immediate connection to the real world is lost and a new problem arises in convincing students that simulated data correctly models the real devices. Also, fitting a simple model to a detailed model is not the same thing as fitting a simple model to measured data. The course evaluation form also revealed that students agree that measurements cannot be replaced by simulated data without a loss of student motivation. Other universities or companies can also use the remote laboratory for characterization of proprietary devices sent to us and, in such cases, simulation software is not an alternative. Of course, one can imagine using prestored data and make the students pretend that they are measuring on real devices. However, this would not merely be an issue of poor teaching but also of ethics.

In the evaluation, we noticed that it is important to prepare a well-structured laboratory experiment manual for the students who are expected to perform the experiments without an instructor at hand. At the same time, it becomes unavoidable for the students to really meet the challenge of understanding instructions where there is no opportunity for hand-holding supervision. However, it is important that instructors are available to answer questions during daytime and that a forum for discussion is open to the students. In the traditional laboratory assignments, there are the opportunities for a supervisor to assist on a just-in-time basis, that is, to be present when the students are confronted by challenges so that rewarding discussions occur in the spur of the moment. Some of these features are lost in I-Lab. However, instead, the introduction of oral presentations offers opportunities to discuss with the students in a situation where the questions have had time to mature, leading to a more rewarding discussion. In our experience, much of the more trivial parts of supervision is better handled among peers within the student community itself. Then student groups that come for teacher assistance using I-Lab have been filtered out efficiently (only a handful sought teacher assistance help out of 175 groups, albeit a few more would have benefited from more extensive supervision). An interesting issue is that ambitious students, no longer hampered by the limitations of a preset lab exercise, now could perform to their full potential given the open task assignment.

We have compared the web-based client to the LabVIEW-based client and noticed that there were both advantages and disadvantages with the two interfaces.

Advantages of the web-based client are that it is easier to use and that no source code can be read on the client machines. There is wide platform support available and no special software needs to be installed, except for the requirement that remote users need a web browser.

An advantage of the LabVIEW-based system is that remote control is much easier to implement, since both server and client are programmed in LabVIEW. It has more flexibility and graphical detail, and security is tighter since only users with access to the client VIs can use the system. The administrator can choose via the server interface which users should be allowed to access the system. A disadvantage is that it requires remote users to have the current version of LabVIEW Player installed or else they need to download the software. Also, remote users may have access to parts of the source code.

6.7 CONCLUSION

We believe that I-Lab enhances student learning and motivates students to undertake more advanced data analysis since less time needs to be spent on practical details. I-Lab brings students in close contact with real devices at the same time as it frees resources for practical experimental training by consuming a minimum of resources in terms of measurement instrumentation, facilities, and supervision.

As more people are given the possibility of sharing the same measurement equipment and controlling it remotely from their offices or even from their homes, expensive equipment will become more cost effective. Interuniversity cooperation also leads to increased availability of advanced measurement instrumentation. Also, advanced university instrumentation can be made available to external organizations and companies can make advanced technology available to students.

Based on this concept, laboratory courses in many disciplines of engineering and science can be offered to students anywhere in the world. I-Lab has been received positively in the introductory microelectronic device course, and we are planning to expand I-Lab and to use it in other courses.

ACKNOWLEDGMENTS

The authors gratefully acknowledge financial support from Nordunet2 (grant 00/1754-24 Internet Technology in Laboratory Modules for Distance-Learning). We would also like to acknowledge fruitful cooperation with professor Tor Fjeldly from Unik, the University Graduate Center at Kjeller, Norway, and Raymond Berntzen, whose staff came down to help us with some of the technical issues in getting I-Lab running.

REFERENCES

J. A. del Alamo, J. Hardison, G. Mishuris, L. Brooks, C. McLean, V. Chan, and L. Hui, "Educational Experiments with an Online Microelectronics Characterization Laboratory," in *Proceedings of the International Conference on Engineering Education*, Manchester, United Kingdom, pp. O102.1–7, 2002. See also www-mtl.mit.edu/users/alamo/weblab/ index.html.

Distance Learning Solution Guide, National Instruments, www.ni.com/academic/dist_publications.htm.

T. A. Fjeldy, J. O. Strandman, and R. Berntzen, "LAB-on-WEB—A Comprehensive Electronic Device Laboratory on a Chip Accessible via Internet," in *Proceedings of the International Conference on Engineering Education*, Manchester, United Kingdom, pp. O337.1–5, 2002. See also www.lab-on-web.com/.

D. Geoffroy, T. Zimmer, M. Billaud, Y. Danto, H. Effinger, W. Seifert, J. Martinez, and G. Francisco, "A Practical Course in a Virtual Lab," in *Proceedings of the 12th EAEEIE Conference*, Nancy, France, 2001. See also www.retwine.net/.

H. Shen, Z. Xu, B. Dalager, V. Kristiansen, Ø. Strøm, M. S. Shur, T. A. Fjeldly, J. Lü, and T. Ytterdal, "Conducting Laboratory Experiments over the Internet," *IEEE Trans. Ed.*, Vol. 42, No. 3, pp. 180–185 (1999). See also `nina.ecse.rpi.edu/ shur/remote/`.

A. Söderlund, F. Ingvarson, P. Lundgren, and K. O. Jeppson, "The Remote Laboratory—A New Complement in Engineering Education," in *Proceedings of the International Conference on Engineering Education*, Manchester, United Kingdom, 2002. See also `www.ic.chalmers.se/ilab/`.

INDEX

Lab on the Web: Running Real Electronics Experiments via the Internet
Edited by Tor A. Fjeldly and Michael S. Shur
ISBN 0-471-41375-5   Copyright © 2003 John Wiley & Sons, Inc.